EPA/600/R-14/009

Avian Life-History Profiles for Use in the Markov Chain Nest Productivity Model (MCnest)

Version: 12 December 2013

Richard S. Bennett and Matthew A. Etterson

Mid-Continent Ecology Division
National Health and Environmental Effects Research Laboratory
Office of Research and Development
U. S. Environmental Protection Agency
Duluth, MN 55804

Table of Contents

Acknowledgments – We wish to thank Dr. Jill Awkerman and Elyssa Arnold for their review of the process for developing and documenting the default species profiles presented in the 11 December 2013 edition of the Species Library for the basic version of MCnest. We also wish to thank Mary Ann Starus for the editorial review of this report.

Introduction

The Markov Chain nest productivity model, or MCnest, quantitatively estimates the effects of pesticides or other toxic chemicals on annual reproductive success of avian species (Bennett and Etterson 2013, Etterson and Bennett 2013). The Basic Version of MCnest was developed as a generalized model requiring a limited number of life-history parameters that can be applied to a broad range of species with limited life-history data. Also, the model treats avian nesting in a manner consistent with the way in which nesting parameters are typically estimated in the field, thus reducing the potential for currency mismatch between the model and available ecological data for the species of interest. The suite of life-history parameters is used to describe the typical series of events during a breeding season of a species. The database of life-history parameters is being developed specifically for use in the Basic Version of MCnest, and the content of the database reflects the specific needs of the model.

This document describes the approach used in creating a database of default life-history profiles for various avian species and an initial list of life-history profiles. The goal is to increase consistency among model users by providing a database of default life-history profiles developed using consistent approaches for selecting parameter values. The MCnest user can choose a default life-history profile from the list of species available in the database, modify an existing profile, or create a new species profile by providing all of the input parameter values.

At present, both MCnest and the species profile database focus primarily on altricial species (i.e., species with chicks characterized as hatching with eyes closed, with little down, incapable of immediately departing the nest, and fed by parents); however, a few precocial species (i.e., species with chicks characterized as hatching with eyes open, downy, and capable of leaving the nest soon after hatch) have been included. The current suite of species represents a range of life-history strategies. It is recognized that as the MCnest model evolves in the future, it may become necessary to modify the contents of the species profile database.

Approach

The species profile database is based on information found in existing literature. The primary literature source for determining values for life-history parameters is The Birds of North America (BNA) series consisting of 716 species accounts written by experts on each species (Available from: http://bna.birds.cornell.edu/bna/). If a BNA species account lacks information on specific parameters or additional information is needed to clarify information on a parameter, additional published literature from journal articles, books, and/or reports is reviewed.

The goal for each required life-history parameter is to determine a single typical value that is representative of that parameter for that species. It is recognized that there is variation in the values for each of the parameters among individuals at a specific time and location, among locations, among breeding seasons, and among techniques and study designs used to collect data. Also, some parameters may vary over the course of a breeding season within a population (e.g., clutch size, daily nest mortality rates). Despite this variation in parameter estimates, we have chosen in the Basic Version of MCnest to select a single typical value rather than try to describe a distribution of values for each parameter. There are several reasons for this decision. First, for most species, there is insufficient information to understand the extent to which factors such as location, season, and study methods truly affect the distribution of parameter values for a species. Many studies do not report sufficient information to define a distribution (e.g., they report a range, the mode, or the earliest egg laying date), and if they do report a mean and variance term, there typically is no attempt to determine the appropriate distribution for the dates. Also, most studies report on observations at a point in space and time, but it may be unclear to what extent those observations represent values for the species as a whole. Consequently, describing parameters as a distribution would require many additional assumptions based on limited or no empirical data. Second, the two parameters that describe the start and end of egg laying can vary significantly with latitude, altitude, or other geographic attributes. If these parameters were to be described as distributions, it would be more appropriate to do it on the basis of a specific location or region rather than for the entire range of the species, because at any specific location the range of possible dates would be narrower than for the species as a whole. Third, describing the life-history parameters as a distribution rather than a single typical value would increase the complexity of the MCnest model significantly and increase the runtime for each simulation. If the simplicity of defining species profiles using single values was the greatest limitation in using MCnest to assess the impact of pesticide exposure on avian productivity, the greater model complexity may be a worthwhile tradeoff. However, the greatest limitation in using MCnest for addressing the specific questions related to annual reproductive success is the quality of the avian toxicity test results and the pesticide exposure estimation (See Bennett and Etterson [2006, 2013] for further discussion). Consequently, we have concluded that without first addressing some of the limitations in the assessment of pesticide effects, increasing the complexity of the breeding season ecology portion of the model by integrating additional stochasticity would not greatly improve model performance as a risk assessment tool. However, this is a topic that needs to be reevaluated as the model evolves in future versions.

The approach for selecting a typical value for each life-history parameter cannot be precisely defined. Reporting of information in the literature is highly variable. For some parameters, authors vary in the way they report information so that it is necessary to make a judgment on how best to interpret the information. For example, there is no standard format for authors reporting the starting and ending dates of egg laying, with some reporting only extreme dates, some reporting the mode or peak of laying, and others reporting information on the distribution of egg laying dates. In general, the typical value reflects some measure of central tendency of available data. For species with several published studies with consistent results, a mean value may be selected (or a weighted mean where sample sizes are also reported), while for other species the selected values may come from a single high quality study, even if other lower quality studies report relevant information. For some parameters, such as the waiting periods after a nest success or failure prior to starting a new nest attempt, there may not be specific quantitative data, but there may be sufficient qualitative information on which to base a reasonable estimate of a typical value. One constraint on the selection of typical values is that all of the parameters reflecting time durations (except the inter-egg laying interval) must be expressed as integers because MCnest uses a daily time step.

The species profile database captures three categories of information: 1) a suite of life-history parameters required for MCnest estimations of productivity; 2) information on female body weight and diet composition during the breeding season required for converting pesticide application rates to initial dietary doses; and 3) information on field estimates of reproductive success, when available, for comparison with MCnest results. The parameters required by

MCnest are presented in the database as single values typical for the species, along with a narrative describing the supporting data, references, and rationale for selecting that value.

MCnest uses the suite of life-history parameters to simulate a population of females as they progress through a breeding season. Parameters such as the initiation probability and the daily nest mortality rates introduce asynchrony in the timing of nesting attempts within the population. As each female transitions from one breeding phase to the next, MCnest tallies the number of nest attempts and the number of successful broods. For more discussion on the decision points guiding transitions between breeding stages and on interpretation of the graphical representation of the population transitioning through breeding phases during a season (i.e., the phase diagram), see the Technical Manual for the Basic Version of MCnest (Bennett and Etterson 2013).

Once typical values are selected for the entire suite of parameters required by MCnest, the values are loaded into MCnest, and a no-pesticide simulation is run to evaluate the overall performance in describing the breeding season of that species. Model outputs, such as the mean number of successful broods or fledglings per female per year and the mean number of nest attempts per female per year, can be compared to literature values to evaluate the performance of a species profile in MCnest. Etterson et al. (2009) discuss some of the issues affecting the comparison of MCnest simulations to empirical data sets. However, when comparing MCnest outputs to literature values, it is important to recognize that literature-derived information on reproductive success can be limited in scope with no clear understanding of how representative that information is of the species as a whole.

Required parameters

The specific list of life-history parameters required for MCnest includes:

- Initiation probability
- Daily background nest failure rate during laying and incubation (m_1)
- Daily background nest failure rate during nestling rearing (m_2)
- Date of first egg laid in first nest of season (T_1)
- Date of first egg laid in last nest of season (T_{last})
- Length of rapid follicle growth period in days for each egg (rfg)
- Length of eggshell formation period in days for each egg (ef)
- Mean clutch size (*clutch*)
- Mean inter-egg laying interval in days (eli)
- Egg on which female typically begins incubation – penultimate vs ultimate (penult)
- Duration in days from start of incubation to hatch (I)
- Duration in days from hatch to fledging of nestlings (N)
- Duration in days since nest failure due to non-pesticide reasons until female initiates new nest (W_e)
- Duration in days since nest failure due to pesticides until female initiates new nest (W_p)
- Duration in days since successful fledging until female initiates new nest (W_f)
- Mean female body weight (*BdyWt*)
- Diet composition of adults and juveniles during breeding season

- Mean number of fledglings per successful nest (*fpsn*)

The definition for each life-history parameter and the rationale used in selecting the typical value for each parameter is discussed in the following sections:

a. Initiation probability

Although species vary in the degree to which the initiation of first nests is synchronized, there is variation in nest initiation dates among females in a population. The initiation probability determines the timing of the first nest attempt for each female in a model simulation. Starting on the earliest date for the first egg laid in the first nest of a season (T_1), the initiation probability defines the daily probability of initiating the first nesting attempt for each female. This parameter is used to create a distribution of nest initiation dates because birds in the wild do not all lay their first egg on the same date. The current default value for the initiation probability in each species life-history profile is 0.25 (i.e., each day 25% of the remaining females in the population initiate their first nest attempt). An initiation probability of 0.25 results in approximately 75% of females starting their first nest in the first 5 d of the breeding season and 95% within 10 d. For most species, there is insufficient information to empirically define a distribution of first nest initiations, so this parameter is used to create a distribution of initiation dates for each simulation. The initiation probability must be > 0 and ≤ 1. If the initiation probability is set equal to 1, all females initiate nests on the same date (i.e., the first day of egg laying, T_1). The lower the initiation probability, the broader the distribution of nest initiation dates.

b. Daily background nest failure rate during laying and incubation (m_1)

Nests fail due to a variety of non-pesticide causes, such as predation and adverse weather events. Many nesting studies report data on nest survival during laying and incubation as a daily nest survival rate (s_1) using methods, such as those described by Mayfield (1961, 1975), to account for biases existing when the fate of each nest is not known throughout the entire period. The daily nest mortality rate (m_1) is the complement of the daily survival rate (i.e., $m_1=1-s_1$).

Other studies may report only the apparent nest survival or failure rate during the egg laying and incubation periods (i.e., # successful or failed nests/total # nests). The apparent rates can be converted to the daily rate over the number of days for the egg laying and incubation periods, as follows:

$$s_1 = \sqrt[a_1]{S_1} = \sqrt[a_1]{1 - M_1} .$$

where S_1 is overall survival rate for egg laying and incubation periods, M_1 is the overall failure rate for egg laying and incubation periods, and a_1 is the age, in days since the first egg was laid, at which nests typically hatch. However, this approach introduces bias by assuming knowledge of the fate of nests over the entire period when this is not the case. Apparent nest success rates overestimate the daily nest survival rates because some nests may fail before detection; however, the degree of overestimation varies considerably due to the specific methods used in studies.

Some studies will report only an overall apparent nest survival rate for the entire nesting period (i.e., egg and nestling phases). As in other situations, overall nest survival rates can be converted to the daily rate over the number of days for the entire nest period (i.e., egg laying, incubation, and nestling-rearing periods), with the same daily rate assigned to m_1 and m_2.

c. Daily background nest failure rate during nestling rearing (m_2)

Many nesting studies report data on daily nest survival rates (s_2) during the nestling-rearing phase. Nest survival is defined as one or more nestlings surviving to fledge from the nest. The daily nest mortality rate during nestling rearing (m_2) is the complement of the daily survival rate ($m_2 = 1 - s_2$). When only an apparent nest survival rate for the brood-rearing phase is reported, it can be converted to a daily nest survival rate over the number of days in the brood-rearing phase, following methods described in the previous section.

For precocial species that depart the nest shortly after hatching, s_2 represents the probability that one or more juveniles in a brood survives to the age of fledging. The definition for fledging is less precise for precocial species and is typically related to when the juveniles have reached a certain level of independence, such as achieving flight or the break-up of the brood. Many studies that report brood survival rates will also define an age that represents fledging. For example, in the mallard profile, the brood survival rate data is based on survival of ducklings in a brood to 60 d post-hatch. Consequently, since the age at which a species fledges is to some degree defined by the authors of studies that measure brood survival, there may be differences among precocial species in the degree of development at the specified time of fledging.

d. Date of first egg laid in first nest of season (T_1)

The length of the clutch initiation period is defined by the difference between the first egg in the first and last nests of the season (i.e., $T_{last} - T_1$). In the Basic Version of MCnest, new nests can only be initiated within this period. However, even though some literature sources report extreme egg laying dates, the T_1 parameter is intended to represent when egg laying typically begins in the first nest attempt for the species of concern. Using extreme laying dates for T_1 and T_{last} may overestimate the length of the clutch initiation period for a species, resulting in more nest attempts than are commonly observed. The value for T_1 could represent the mean date reported from multiple studies or it may represent a mean date reported in a high quality study over multiple years.

In the Species Library file in MCnest (i.e., SpeciesLibrary.xlsx), T_1 is presented as two parameters for the month (i.e., initial month or "imn") and day (i.e., initial day or "idy") of the first egg in the first nest.

e. Date of first egg laid in last nest of season (T_{last})

Similar to T_1, the parameter T_{last} represents the typical date for the first egg in the last nest of the season. Again, this is not intended to represent extreme egg laying dates, as that would serve to extend the simulated length of the breeding season and bias the model output.

In the Species Library file in MCnest (i.e., SpeciesLibrary.xlsx), T_{last} is presented as two parameters for the month (i.e., end month or "emn") and day (i.e., end day or "edy") of the first egg in the last nest.

f. Length of rapid follicle growth period in days for each egg (rfg)

Unlike fish, amphibians, and reptiles, birds do not lay their eggs in masses. Instead, most birds lay an egg each day until they complete a clutch, while some birds may lay an egg every other day or some other period longer than one day. The follicles that develop into egg yolks also start growing on a staggered schedule over a several day period, known as the rapid follicle growth (*rfg*) period. During the *rfg* period yolk material is deposited to the growing follicle until it reaches the size of a fully formed yolk just prior to ovulation.

Although estimates for the duration of the *rfg* period are available only for a subset of species, there is sufficient empirical information for estimating the *rfg* period for most species. Alisauskas and Ankney (1992) developed an allometric equation for estimating the duration of the *rfg* period in birds. Many song birds (i.e., passerines) have an *rfg* period of 3 to 4 d while in doves and pigeons the period is approximately 6 d. See the MCnest Technical Manual Appendix A (Estimating the length of the rapid follicle growth period) for additional detail. The value selected for the duration of the *rfg* period must be an integer.

g. Length of eggshell formation period in days for each egg (ef)

After the yolky follicle is ovulated, it enters the oviduct for deposition of the albumin followed by formation of the egg membrane and shell. This process usually takes approximately 24 hours for most species. Consequently, the eggshell formation period is fixed at 1 d in the model, and thus does not show up in MCnest as one of the input parameters.

h. Mean clutch size (clutch)

Clutch size is one of the most commonly reported avian life history parameters in the literature. Here again, the intent is to select a value representing the typical clutch size of a species rather than extreme values. The value selected for *mean clutch size* must be an integer.

i. Mean inter-egg laying interval in days (eli)

As mentioned above, most birds lay one egg each day, while other species may have a longer mean inter-egg laying interval. The value for *eli* can be any value ≥ 1 day and can be expressed as a decimal value.

j. Egg on which female typically begins incubation – penultimate vs ultimate (penult)

Those species beginning incubation after the last egg is laid are assigned a value for *penult* of 0, while those beginning with the penultimate egg are assigned a value of 1. For those species where both options are possible, a judgment is made as to which option is more typical for the species.

10

k. Duration in days from start of incubation to hatch (I)

The duration of the incubation period (I) also is a commonly reported life-history parameter in the literature. A typical value for the duration of the incubation period should be selected and it must be expressed as an integer. For pesticides that affect egg hatchability because of embryotoxicity due to *in ovo* exposure or infertility, there is a related parameter known as "doomed incubation" or Id. When pesticide exposure is high enough to cause embryotoxicity or reduced fertility, it is assumed the female does not become aware that these effects have occurred until the time at which eggs are expected to hatch. Consequently, the female continues to incubate the clutch of eggs that has failed a decision point and is considered to be "doomed," but the nest attempt does not actually fail until the end of incubation period when the eggs fail to hatch on schedule. In the Basic Version of MCnest the duration of the Id period is set to the same value as I.

l. Duration in days from hatch to fledging of nestlings (N)

The duration of the brood-rearing period also is a commonly reported life-history parameter in the literature. A typical value for the duration of brood rearing should be selected and it must be expressed as an integer. Nestlings of some species can leave the nest early when stressed by predators or weather events such as floods. While many studies report a wide range of fledging durations that reflect that some nests fledge early under stress, the intent is to select a typical value for the brood-rearing period reflective of non-stressed conditions (i.e., how long do nestlings typically remain in the nest if not stressed?).

For precocial species, the definition for fledging is less precise and is typically related to when the juveniles have reached a certain level of independence, such as achieving flight or the break-up of the brood. For species profiles used in the Basic Version of MCnest, the duration of N may be linked to available data for parameterizing the daily brood mortality rate (m_2). Authors may vary in how they describe fledging in precocial species, but if the value for m_2 is derived from an estimate of brood survival up to a specific number of days post-hatch, the duration of N could be defined by that same study.

m. Duration in days since nest failure due to non-pesticide reasons until female initiates new nest (W_e)

After a nest failure due to environmental causes such as predation or weather, females may attempt to renest after a period of recovery and reinitiation of the egg formation process. The value for W_e represents the duration from nest failure until the first egg is laid in a new nest and must be expressed as an integer. Some high quality studies have data on the duration of this period. However, for many species, there may not be specific quantitative data for parameterizing W_e, and it may be necessary to estimate a value based on other available information. Studies may report that renesting after failure is common and that it occurs rapidly. We assume that the shortest duration for W_e for a female that was incubating or rearing nestlings is the period of time required to form a new egg (i.e., period of rapid follicle growth plus one day for eggshell formation). For some species it also is necessary to factor in sufficient time for building a new nest and/or finding a new mate. For those studies that report specific durations for W_e, it is assumed that the need for nest building and mate finding have been incorporated into the estimates.

n. Duration in days since nest failure due to pesticides until female initiates new nest (W_p)

After a nest failure due to pesticide exposure, females also may attempt to renest after a period of recovery and reinitiation of the egg formation process. W_p represents the duration from nest failure until the first egg is laid in a new nest and must be expressed as an integer. Of all the life-history parameters, W_p may have the poorest amount of information for selecting a value. Pesticide field studies typically do not provide information on the probability or timing of renesting after a pesticide-related nest failure. Occasionally, laboratory reproduction studies are designed to include a period of untreated food at the end of the treatment period. These studies can provide information on the potential for egg production to increase or restart after treatment ends, though it is not clear if this is indicative of the potential for free-ranging birds to renest. Depending on the nature of the pesticide, the model user might assume that W_p equals W_e if birds recover quickly from an exposure. However, for chemicals with prolonged or delayed effects after exposure, a longer duration may be appropriate for W_p. A conservative assumption would be that females do not renest after pesticide failure (i.e., set W_p to a value larger than the length of the breeding season). However, suitable field examples of renesting periods after a pesticide-induced failure have not been found that provide a basis for additional guidance. The W_p cannot be shorter than W_e.

As a default in the Basic Version of MCnest, W_p is set equal to W_e. If a model user decides to change the value for W_p, it is unlikely that there would be sufficient information to set species-specific values, so the value for W_p is not located on the "Life History" window of each species. If the model user is running simulations on a single species, the value for W_p can be changed on the "Set Pesticide" window; however, if multiple species are being simulated using the "Batch mode," there is a toggle switch for overriding the W_p value used for all selected species. If a model user inserts a value that overrides the default W_p values, this value will be used as the W_p for all species, except for species where the W_e value is larger, in which case the value of W_e is used for W_p.

o. Duration in days since successful fledging until female initiates new nest (W_f)

After successfully fledging an initial brood, some species will attempt one or more additional broods. W_f represents the duration from successful fledging until the first egg is laid in a new nest and must be expressed as an integer. In some species, fledglings become the responsibility of the male while the female immediately initiates a new clutch of eggs. In other species, both males and females or the females alone continue to feed and/or care for fledglings until they become independent–a period of up to several weeks–after which the female may become available to start a new nest if time remains in the breeding season. When there is a period of female involvement in post-fledging care, the estimates of the period of time until the female renests found in the literature can be quite variable. It is not always clear if the shorter estimates reflect that some females renest relatively rapidly, even if they still are assisting with fledgling care, or that some females have lost their entire broods prior to becoming independent.

p. Mean female body weight and diet composition during breeding season

MCnest simulations involving a pesticide exposure require information on the mean female body weight (in grams) and diet composition, ideally representing weights and diet during the breeding season, as well as diet composition of juveniles prior to fledging. The body weight and diet information is used in converting application rates into the estimated daily dietary dose (in mg/kg body weight/day) for each species based on the algorithm used in OPP's T-REX model (http://www.epa.gov/oppefed1/models/terrestrial/trex/t_rex_user_guide.htm).

The diet composition in MCnest species' profiles is expressed as the proportion of the diet in each of the six food categories used in T-REX, including 1) short grass, 2) tall grass, 3) broadleaf forage plants, 4) fruits, 5) seeds and pods, and 6) insects and other invertebrates. It is intended that the proportions are based on the mass (wet weight) of each type, but in some species literature information on diet composition may only be expressed as volume or counts of food items. Diet composition changes for many species throughout the year. It also may vary between sexes during the breeding season. For example, females may increase the percentage of invertebrates consumed during egg laying. It is preferable to use diet information for females during the breeding season, when available. The diet composition for nestlings also may change somewhat as hatchlings grow to become fledglings. For example, mourning dove hatchlings start on a diet of crop milk produced by the parents but the diet gradually switches to all seeds by the time of fledging. The goal is to use available information of juvenile diet composition to estimate the percentage of various food type categories in the diet throughout the nestling period.

Female body weights vary through the different phases of the breeding cycle, and literature sources vary widely in the specificity of reporting body weights as a function of seasons, so it is difficult to consistently find mean body weight data linked to a specific phase of the breeding cycle. Ideally, this parameter would represent mean female body weight during the peak of the nesting period. If this is not specifically available, mean weights from a more broadly defined breeding season are acceptable, or lacking that, mean weights throughout the year are acceptable. For some species, body weight data may be available for different subspecies. Depending on the species, the most appropriate "typical" value to use may be from the subspecies predominantly found in agricultural areas or it may be an average of multiple subspecies mean values.

q. Mean number of fledglings per successful nest (fpsn)

Each species profile contains an estimate of the mean number of fledglings per successful nest, which is multiplied by the number of successful nests per female per season (i.e., the primary output from MCnest simulations) to estimate the number of fledglings per female per season. Many studies report the mean number of fledglings per nest, but for this parameter it is important to focus on information expressed as the mean number of fledglings per successful nest, where "successful nest" is defined as a nest producing at least one fledgling. Consequently, the mean number of fledglings per successful nest must be between one and the mean clutch size. Sometimes it is unclear in study reports whether the calculated number of fledglings refers to only those from successful nests or from all nests. Where this cannot be reasonably determined, this information should be used with great caution in determining an estimate of the mean number of fledglings. Sometimes, even though a study does not specifically report the mean number of fledglings per successful nest, it may provide sufficient information to calculate it. Even if there is insufficient data available in the literature for estimating the number of fledglings per successful nest, the MCnest profile requires at least a placeholder value. However, the model user needs to be aware of how the quality of the estimate for number of fledglings per successful nest affects the estimate in MCnest of the mean number of fledglings per female per year. The mean number of fledglings per successful nest does not need to be an integer.

Additional parameters

There are several additional parameters that are not required as part of the species life-history profile, but that are useful for evaluating the performance of MCnest. When information is available on these parameters, it should be captured in the life-history database, at least in narrative form even if it is not possible to determine a numerical estimate. The parameters include:

r. Mean number of nest attempts per female per season

MCnest calculates the mean number of nest attempts (both successful and failed) per female per season as part of the basic output from each model simulation. This parameter is included in the life-history database as a place to summarize information in the literature about the number of nest attempts per female. However, unless a population is closely monitored (e.g., radio-tagged), it may be difficult for investigators to accurately document the number of nest attempts each female makes in the field. Some investigators may report, for example, that females attempt two clutches per year, but this type of qualitative statement may underestimate the number of failed attempts that occur. Consequently, while it is useful to summarize pertinent information in narrative form in the database, it may be very difficult to accurately determine an estimate that reflects the actual number of successful and failed nest attempts per female per season.

s. Mean number of successful broods per female per season

The primary output from MCnest is the mean number of successful broods per female per season (i.e., broods that successfully fledge at least one juvenile). This parameter is included in the life-history database as a place to summarize information in the literature about the number of successful broods per female. However, like the number of nest attempts, unless a population is closely monitored, it may be difficult for investigators to accurately document the number of successful broods per female in the field. Again, some investigators may report, for example, that females *attempt* to raise two broods per season, but this may overestimate the number of successful broods because many females in the population may have produced only one or no *successful* broods during the season. Consequently, while it is useful to summarize pertinent information in narrative form in the database, it may be very difficult to accurately estimate the mean number of successful broods per female per season unless the study is carefully designed for that purpose.

t. Mean number of fledglings per female per season

In birds, the annual reproductive success (i.e. ARS) or fecundity rate often is expressed as the mean number of fledglings per adult female in the population per season. Field estimates of this parameter are only possible from very closely monitored populations because it requires tracking the nesting activity of individual females through the breeding season. Some studies report estimates of this parameter or provide data that can lead to an estimate. This parameter is included in the life-history database as a place to summarize information in the literature about the number of fledglings per female per season. Field estimates are important because they provide a means for evaluating how well MCnest output represents the breeding season characteristics of a species. However, field estimates have their own limitations (e.g., they may only reflect productivity at a given place at a given time), so it is important to assess the representativeness of field-derived productivity estimates when comparing with MCnest outputs. It is useful to summarize pertinent information in narrative form in the database, but it may be very difficult to accurately estimate the mean number of fledglings per female per season unless the study is designed for that purpose.

Species Profile Database

The Species Profile Database (i.e., SpeciesLibrary.xlsx) is anticipated to evolve over time as species are added or revised. The current version of the Species Profile Database (See Appendix A) dated 12 December 2013 contains profiles on the following species listed in taxonomic order:

Canada goose–giant subspecies (*Branta canadensis maxima*)
Mallard (*Anas platyrhynchos*)
Blue-winged teal (*Anas discors*)
Northern bobwhite (*Colinus virginianus*)
American kestrel (*Falco sparverius*)
Killdeer (*Charadrius vociferus*)
White-winged dove (*Zenaida asiatica*)
Mourning dove (*Zenaida macroura*)
Northern flicker (*Colaptes auratus*)
Willow flycatcher (*Empidonax traillii*)
Eastern phoebe (*Sayornis phoebe*)
Ash-throated flycatcher (*Myiarchus cinerascens*)
Eastern kingbird (*Tyrannus tyrannus*)
Blue jay (*Cyanocitta cristata*)
American crow (*Corvus brachyrhynchos*)
Horned lark (*Eremophila alpestris*)
Tree swallow (*Tachycineta bicolor*)
Barn swallow (*Hirundo rustica*)
Carolina chickadee (*Poecile carolinensis*)
Black-capped chickadee (*Poecile atricapillus*)
Verdin (*Auriparus flaviceps*)
Carolina wren (*Thryothorus ludovicianus*)
House wren (*Troglodytes aedon*)
Blue-gray gnatcatcher (*Polioptila caerulea*)
Eastern bluebird (*Sialia sialis*)
Wood thrush (*Hylocichla mustelina*)
American robin (*Turdus migratorius*)
Northern mockingbird (*Mimus polyglottos*)
Cedar waxwing (*Bombycilla cedrorum*)
Ovenbird (*Seiurus aurocapillus*)
Common yellowthroat (*Geothlypis trichas*)
Yellow warbler (*Setophaga petechia*)
Yellow-rumped warbler (*Setophaga coronata*)
Cassin's sparrow (*Peucaea cassinii*)
Chipping sparrow (*Spizella passerina*)
Field sparrow (*Spizella pusilla*)
Vesper sparrow (*Pooecetes gramineus*)
Lark sparrow (*Chondestes grammacus*)
Lark bunting (*Calamospiza melanocorys*)
Savannah sparrow (*Passerculus sandwichensis*)

Grasshopper sparrow (*Ammodramus savannarum*)
White-crowned sparrow–Nuttall's subspecies (*Zonotrichia leucophrys nuttali*)
White-crowned sparrow–Puget Sound subspecies (*Zonotrichia leucophrys pugetensis*)
Dark-eyed junco (*Junco hyemalis*)
Northern cardinal (*Cardinalis cardinalis*)
Dickcissel (*Spiza americana*)
Bobolink (*Dolichonyx oryzivorus*)
Red-winged blackbird (*Agelaius phoeniceus*)
Eastern meadowlark (*Sturnella magna*)
Western meadowlark (*Sturnella neglecta*)
Brewer's blackbird (*Euphagus cyanocephalus*)
Common grackle (*Quiscalus quiscula*)
Boat-tailed grackle (*Quiscalus major*)
Great-tailed grackle (*Quiscalus mexicanus*)
House finch (*Carpodacus mexicanus*)
American goldfinch (*Carduelis tristis*)
House sparrow (*Passer domesticus*)

Two profiles exist for the tree swallow, representing differences in some of the life-history parameters between populations in the northern and southern portions of the species range. These are included as a possible prototype for addressing regional differences in life-history parameters for species with extensive ranges. Also, two profiles each are included for the wood thrush and the ovenbird reflecting the influence of habitat (i.e., fragmented forest vs contiguous forest) on the background daily nest mortality rate and the mean number of fledglings per successful nest.

The 12 December 2013 Species Library includes the species contained in the original 10 January 2013 Species Library plus many new species profiles. There are a few changes to the original profiles. The rapid follicle growth (*rfg*) period for northern flicker, Carolina chickadee, and black-capped chickadee was modified based on the allometric equation in Alisauskas and Ankney (1992), and the end of egg laying date (i.e. (T_{last}) for northern mockingbirds was revised from August 1 to June 30.

Using Default Species Profiles in MCnest

When opening MCnest, the model will access the default species profiles from the SpeciesLibrary.xlsx file. As mentioned above, the model user may use the default species profiles, modify an existing profile, or create new profiles. Each of the life-history parameters is editable in the Life History window, except the value for the waiting period after a pesticide-related nest failure (i.e., *Wp*), which is editable on the Pesticide window or under the "Species" option in the Batch mode text box (see User's manual for more detail). The model user also may create an alternative database (using the same format found in SpeciesLibrary.xlsx) in the MCnest working directory and switch to that database by selecting "Load Species Library" from the drop-down menu under the MCnest tab.

The default species profile database was developed to increase consistency among model users by providing a common set of typical values for the life-history parameters for each species.

However, the typical values selected to represent the species as a whole may not be the best description of a species at a specific location, or for specific assumptions about habitat quality that can affect the daily nest mortality rates. Depending on the needs of a particular assessment, the model user may want to modify species profiles to better fit a specific scenario (e.g., site-specific assessment), if appropriate data are available.

A frequently asked question is whether or not the species profiles are designed to represent worst-case scenarios for risk assessment. In other words, would the default life-history parameters always produce outcomes indicating the highest level of risk or could the selection of different life-history parameter inputs (e.g., different egg laying dates, daily nest mortality rates, etc.) produce outcomes with higher levels of risk? The answer is that the default species profiles are not designed to represent worst-case scenarios and it is not possible to select a set of life-history parameter inputs that universally produces worst-case outcomes. Even if a particular profile was modified to provide a worst-case outcome for a specific chemical and application date(s), under a different type of chemical or application date(s), the outcome could be far from the worst-case outcome. Consequently, the default specifies profiles are intended to represent the typical life-history parameter values for the species as a whole, without regard to how changes in those values affect the perception of risk from pesticide exposure.

References

Alisauskas, R. T. and C. D. Ankney. 1992. The cost of egg laying and its relationship to nutrient reserves in waterfowl. In: Ecology and Management of Breeding Waterfowl. B. Batt (Ed), University of Minnesota Press. Pp 30-61.

Bennett, R. S. and M. A. Etterson. 2006. Estimating pesticide effects on fecundity rates of wild birds using current laboratory reproduction tests. *Human Ecol. Risk Assess.* 12:762-81.

Bennett, R. S. and M. A. Etterson. 2007. Incorporating results of avian toxicity tests into a model of annual reproductive success. *Integr. Environ. Assess. Manage.* 3(4):498-507.

Bennett, R. S. and M. A. Etterson. 2013. Technical Manual for Basic Version of the Markov Chain Nest Productivity Model (MCnest). U.S. Environmental Protection Agency, Office of Research and Development. U.S. EPA 600/R-13/033.

Etterson, M. A., R. S. Bennett, E. L. Kershner and J. W. Walk. 2009. Markov chain estimation of avian seasonal fecundity. *Ecol. Appl.* 19(3):622-630.

Etterson, M. A. and R. S. Bennett. 2013. Quantifying the effects of pesticide exposure on annual reproductive success of birds. *Integr. Environ. Assess. Manage.* 9:590-599.

Mayfield, H. 1961. Nesting success calculated from exposure. *Wilson Bull.* 73:255-61.

Mayfield, H. F. 1975. Suggestions for calculating nest success. *Wilson Bull.* 87:456-66.

Appendix A. Typical values for life history parameters and the rationale for their selection for each species in the Species Profile Database.

Canada goose–giant subspecies (*Branta canadensis maxima*)			Four-letter Alpha Code: CANG
Species life-history parameters	Model Code	Typical value	Rationale
daily mortality rate during laying & incubation	m1	0.012	"Nest success varies greatly among populations, likely reflecting variation in predator abundance or availability of secure nest sites. Success (proportion of nests that hatch at least 1 egg) varies from <20% (Bromley and Jarvis 1993, CRE) to ≥85% (Hilley1976, Bromley and Jarvis 1993)" as quoted from Mowbray et al. (2002). Brakhage (Table 13: 1965) compared nest success at 14 sites ranging from 24% to 84%, with mean of 64%. Assuming a typical apparent nest success rate at hatching of 64% over 37 d (i.e., 28+(6*1.5)), the daily nest mortality rate would be 0.012.
daily mortality rate during nestling-rearing	m2	0.0026	"Young leave breeding areas with parents; in large subspecies, offspring remain with parents throughout first year" as quoted from Mowbray et al. (2002). Documenting brood survival rates is complicated by the formation of crèches where multiple broods merge and are accompanied by multiple pairs of adults. "If 85% of older geese nested and 75% of these hatched at least 1 young, then about 64% of older adults started with a brood and 58% still had a brood at fledging. Thus, their success in rearing at least 1 young from hatching to fledging was 91%" as quoted from Raveling (1981). Using this reasoning, Raveling also reports 47% of 2 yr old geese nest, so with a 75% nest success rate, 35% started with a brood and 25% still had a brood at fledging, so brood success for 2 yr olds was approximately 70%. Since most breeding adults are over 2 yr old, assume that an overall brood success rate of 85% over a 63 d period from hatch to fledging results in a daily brood mortality rate of 0.0026.
date of first egg of first nest (dd-mmm)	T1	Mar-20	"Timing of egg-laying varies considerably, from as early as mid-to- late Mar in *B. c. maxima* and *B. c. moffitti* to as late as mid-Jun in *B. c. canadensis*, *B. c. parvipes*, *B. c. hutchinsii*, and *B. c. minima*" as quoted from Mowbray et al. (2002). In n.w. MO, first eggs were laid by giant Canada geese between March 15 and 20 each year of 3 yr study and 94% of nests were initiated within 4 weeks of start of laying (Brakhage 1965). Assume typical egg laying starts March 20 and ends April 20.
date of first egg of last nest (dd-mmm)	Tlast	Apr-20	
length of rapid follicle growth period (RFG) for each egg (days)	rfg	13	13 d (Alisauskas and Ankney 1992)
mean clutch size	clutch	6	"Usually 2–8; generally fewer eggs in late nests (Raveling and Lumsden 1977, Rohwer and Eisenhauer 1989), in years of high population density (Hanson and Eberhardt 1971), and among younger birds (Hofman 1982, Aldrich and Raveling 1983, CRE). Typical values for *B. c. maxima*, 5.6 (Manitoba; Cooper 1978), 5.6 (Missouri; Brakhage 1965)" as quoted from Mowbray et al. (2002).
mean intra-egg laying interval (days)	eli	1.5	"Eggs generally laid at 35-h intervals (range 30–40; Kossack 1950, Cooper 1978, Carriere et al. 1999), with interval shorter for late eggs" as quoted from Mowbray et al. (2002).

Description	Code	Value	Notes
egg on which female typically begins incubation—penultimate (1) or last (0)	penult	0	"Nest attendance gradually increases during laying period until penultimate egg laid (Cooper 1978). Continuous incubation likely starts after last egg in small clutches, earlier in large clutches" as quoted from Mowbray et al. (2002).
duration from start of incubation to hatch (days)	I	28	"Varies with body size and breeding latitude of subspecies; 25 d in *B. c. minima* (CRE), 26–27 d in *B. c. hutchinsii* (Jarvis and Bromley 2000), and 28 d in *B. c. occidentalis* (Bromley 1998) and *B. c. maxima* (Cooper 1978)" as quoted from Mowbray et al. (2002).
duration from hatch to fledging of nestlings (days)	N	63	"Fledging requires 7–8 wk for *B. c. moffitti* in California (Moffitt 1931); 9 wk after hatching for both *B. c. interior* at James Bay, Ontario (Hanson 1965), and *B. c. maxima* in Manitoba (Balham 1954)" as quoted from Mowbray et al. (2002).
duration since nest failure due to other reasons until female initiates new nest (days)	We	16	"Renesting 14–20 d following nest loss in some lower latitude populations (Atwater 1959, Reinecker and Anderson 1960, Bromley 1998)" as quoted from Mowbray et al. (2002). In n.w. MO, "10 of 30 pairs that lost clutches during laying continued to lay in new sites nearby, 7 renested after an interval of 5 to 22 d depending upon the number of eggs laid, and 13 made no further attempt to lay (Brakhage 1965)" as quoted in Bellrose (1976). Assume 16 d is a typical duration for We.
duration since successful fledging until female initiates new nest (days)	Wf	16	"Single-brooded but for a few exceptions in southernmost breeding populations (e.g., Brakhage 1985)" as quoted from Mowbray et al. (2002). Given the long period of juvenile care and the short window for egg laying, renesting after success is not usually possible, and the value for Wf will not influence model outcomes. Assume 16 d as a placeholder in the profile.
female body weight (g) during breeding season	BdyWt	4825	*B. c. maxima*: Females: 4,825.0 g ± 425.1 (n=10) and males: 4,858.3 g ± 280.3 (n=6) (Coluccy 2001).
diet composition during breeding season	H		"Canada Geese depend primarily on grasses, sedges, or other green monocots during periods of increase in lean body mass, primarily the growth period in summer (Sedinger and Raveling 1984) and spring premigration and migration periods (McLandress and Raveling 1981a, Coleman and Boag 1987a). Graminoids represent >90% of summer diets of resident *B. c. maxima* broods along Atlantic coast (Buchsbaum and Valiela 1987). In 2 subspecies (*B. c. interior* and *B. c. minima*), goslings almost exclusively eat green leaves of graminoids (Sedinger and Raveling 1984, Bruggink et al. 1994)" as quoted from Mowbray et al. (2002). Although diet varies with location and food availability, assume both adults and juveniles consume 100% short grass.
mean number of nest attempts/ female/ season			
mean number of successful broods/ female/ season			
mean number of fledglings/ successful nest	fpsn	3.44	Based on data in Table 2 of Raveling (1981) for year 1971 and 1972 (i.e., only years that span all age cohorts), calculated a weighted mean of 3.44 fledglings/successfully fledged brood.

mean number of fledglings/ female/ season (ARS)	ARS	1.21	Based on data in Table 2 of Raveling (1981) for year 1971 and 1972 (i.e., only years that span all age cohorts), calculated a weighted mean of 1.21 fledglings/female in the fall census population, which may underestimate the number of juveniles alive at time of fledgling.

Reference List

Aldrich, T. W. and D. G. Raveling. 1983. Effects of experience and body weight on incubation behavior of Canada Geese. *Auk* 100:670-679.

Alisauskas, R. T., and C. D. Ankney. 1992. The cost of egg laying and its relationship to nutrient reserves in waterfowl. In: *Ecology and Management of Breeding Waterfowl*. B. Batt (Ed), University of Minnesota Press. Pp 30-61.

Atwater, M. G. 1959. A study of renesting in Canada Geese in Montana. *J. Wildl. Manage.* 23:91-97.

Balham, R. W. 1954. The behavior of the Canada Goose (*Branta canadensis*) in Manitoba. Phd Thesis. Univ. of Missouri, Columbia.

Brakhage, G. K. 1965. Biology and behavior of tub-nesting Canada Geese. *J. Wildl. Manage.* 29:751-771.

Bromley, R. G. 1998. Conservation assessment for the Dusky Canada Goose (*Branta canadensis occidentalis*). Unpubl. rep. to the Pacific Flyway Council.

Bromley, R. G. and R. L. Jarvis. 1993. The energetics of migration and reproduction of Dusky Canada Geese. *Condor* 95:193-210.

Bruggink, J. G., T. C. Tacha, J. C. Davies, and K. F. Abraham. 1994. Nesting and brood-rearing ecology of Mississippi Valley Population Canada Geese. *Wildl. Monogr.* 126.

Buchsbaum, R. and I. Valiela. 1987. Variability in the chemistry of estuarine plants and its effect on feeding by Canada Geese. *Oecologia* 73:146-153.

Carriere, S., R. G. Bromley, and G. Gauthier. 1999. Comparative spring habitat and food use by two arctic nesting geese. *Wilson Bull.* 111:166-180.

Coleman, T. S. and D. A. Boag. 1987a. Canada Goose foods: their significance to weight gain. *Wildfowl* 38:82-88.

Cooper, J. A. 1978. The history and breeding biology of the Canada Geese of Marshy Point, Manitoba. *Wildl. Monogr.*. 61.

Hanson, H. C. 1965. The Giant Canada Goose. Southern Illinois University Press, Carbondale.

Hanson, W. C. and L. L. Eberhardt. 1971. The Columbia River Canada Goose Population 1950-1970. *Wildl. Monogr.* 28.

Hilley, J. D. 1976. Productivity of a resident Giant Canada Goose flock in northeastern South Dakota. Master's Thesis. South Dakota State Univ. Brookings.

Hofman, D. E. 1982. Breeding experience as a factor influencing clutch size and fertility in captive Canada Geese. *Wildl. Soc. Bull.* 10:384-386.

Jarvis, R. L. and R. G. Bromley. 2000. Incubation behavior of Richardson's Canada Geese on Victoria Island, Nunavut, Canada. Pages 59-64 *in* Towards conservation of the diversity of Canada Geese (*Branta canadensis*). (Dickson, K., Ed.) *Can. Wildl. Serv. Occas. Pap.* no. 103.

Kossack, C. W. 1950. Breeding habits of the Canada Goose under refuge conditions. *Am. Midl. Nat.* 43:627-649.

Mclandress, M. R. and D. G. Raveling. 1981b. Changes in diet and body composition of Canada Geese before spring migration. *Auk* 98:65-79.

Moffit, J. 1931. The status of the Canada Goose in California. *Calif. Fish Game* 17:20-26.

Mowbray, Thomas B., Craig R. Ely, James S. Sedinger and Robert E. Trost. 2002. Canada Goose (*Branta canadensis*), The Birds of North America Online (A. Poole, Ed.). Ithaca: Cornell Lab of Ornithology; Retrieved from the Birds of North America Online: http://bna.birds.cornell.edu.bnaproxy.birds.cornell.edu/bna/species/682.

Raveling, D. G. 1981. Survival, experience, and age in relation to breeding success of Canada Geese. *J. Wildl. Manage.* 45:817-829.

Raveling, D. G. and H. G. Lumsden. 1977. Nesting ecology of Canada Geese in the Hudson Bay Lowlands of Ontario: evolution and population regulation. *Fish Wildl. Res. Rep.* no. 98. Ontario Min. Nat. Resour.

Reinecker, W. C. and W. Anderson. 1960. A waterfowl nesting study on Tule Lake and Lower Klamath National Wildlife Refuges, 1957. *Calif. Dep. Fish Game Bull.* 45:481-507.

Rohwer, F. C. and D. I. Eisenhauer. 1989. Egg mass and clutch size relationships in geese, eiders, and swans. *Ornis. Scand.* 20:43-48.

Sedinger, J. S. and D. G. Raveling. 1984. Dietary selectivity in relation to availability and quality of food for goslings of Cackling Geese. *Auk* 101:295-306.

Mallard (*Anas platyrhynchos*)

Four-letter Alpha Code: MALL

Species life-history parameters	Model Code	Typical value	Rationale
daily mortality rate during laying & incubation	m1	0.0513	Nest success rate varies considerably among studies, habitats, years, etc. Cowardin et al. (1985) determined that mallard nest success rate of 15% was needed to maintain population. Over 36 d nest period (i.e., 28+9-1), an overall nest success rate of 15% translates to a daily nest mortality rate of 0.0513.
daily mortality rate during nestling-rearing	m2	0.0063	Broods successfully fledging at least 1 duckling: 74% (20 of 27) in ND (Cowardin et al. 1985), 63% (17 of 27) in MN (Orthmeyer & Ball 1990). Weighted average=68.5%, so daily brood mortality rate over 60 days is 1 - 0.9937 = 0.0063. Cowardin and Johnson (1979) use an estimate of 77% based on a statement in Ball et al (1975) that "failure to account for total-brood loss …would result in overestimating production by about 30%" (i.e., 1.0/1.3 = 0.77).
date of first egg of first nest (dd-mmm)	T1	Apr-15	Dates based on Figure 2 in Drilling et al (2002). "The mallard begins to nest between April 10 and April 30 over vast reaches of its breeding range" (Bellrose 1976). Johnson et al (1987) use July 15 as end of egg laying. Nesting initiations span about 60 d over much of the range, but as much as 80 d in SD (Bellrose 1976).
date of first egg of last nest (dd-mmm)	Tlast	Jul-15	
length of rapid follicle growth period (RFG) for each egg (days)	rfg	6	6 d for mallards (Alisauskas and Ankney 1992)
length of eggshell formation (EF) period for each egg (days)	ef	1	Although not always reported, a typical period of time from ovulation to oviposition is approx. 1 d. We assume EF is 1 d unless other information exists.
mean clutch size	clutch	9	"Average clutch size of 20 North American studies between 1949 and 1957: 8.72 (range of means 5.7-10.6, n=1468: Dzubin and Gollop 1972: Appendix B)" from Drilling et al. (2002).
mean intra-egg laying interval (days)	eli	1	Usually 1 egg laid/day (Drilling et al. 2002).
egg on which female typically begins incubation–penultimate (1) or last (0)	penult	1	Hens increase amount of time on the nest during laying with nest temperatures adequate for embryonic development starting by 6th egg, but no incubation at night until clutch is complete (Drilling et al. 2002).
duration from start of incubation to hatch (days)	I	28	"Average incubation period 28 d, normal range 23-30 (Palmer 1976) from Drilling et al. (2002).
duration from hatch to fledging of nestlings (days)	N	60	Ducklings begin to thermoregulate at 1 d post-hatch. Hens stay with ducklings until they begin to fly at about 50-60 d. In n. Minnesota, hen stayed with brood average of 50.7 d (Ball et al. 1975). Orthmeyer and Ball (1990) monitored brood success to fledging, which they defined as 60 d. Based on studies monitoring brood survival until approx. 60 d, we define fledging as ability to fly at 60 d.

Description	Abbrev.	Value	Notes
duration since nest failure due to other reasons until female initiates new nest (days)	We	10	Mean renesting interval for mallards on unlimited food is 7.1 d (range 5-10), but on limited food range is 6 -24 d with 42% >10 d (Swanson et al. 1986). Mean of 12.5 d (range 4-15 d) in VT (Coulter and Miller 1968). Sowls (1955) reported that renesting interval increase with number of days of incubation. For females that lose broods renesting interval may be longer. "Wild mallards rarely renest after brood loss, but urban mallards do so frequently (Figley and VanDruff 1982)" from Drilling et al. (2002). Assume mean of 10 d for renesting interval.
duration since successful fledging until female initiates new nest (days)	Wf	10	Wild mallards rarely raise second broods, but some birds in urban and unnaturally crowded populations may (Drilling et al. 2002). If a female does renest after success, interval unlikely to be shorter than We. Since the period of time required to raise a successful brood to fledging exceeds the length of the egg laying window, the selected length of the Wf period does not affect the outcome of the model simulation.
female body weight (g) during breeding season	BdyWt	1200	1197 ± 105 SD spring females in ND (Krapu and Doty 1979), 1200 ± 78 SD prelaying females in ND (Krapu 1981),
diet composition during breeding season	O	O	Breeding females 72% invertebrates and 28% plant material, primarily seeds (Swanson et al. 1985). Ducklings <25 d old eat mostly invertebrates – 90% (Chura 1961), but older ducklings switch to a higher proportion of seeds. Assume juveniles consume 90% invertebrates and 10% seeds.
mean number of nest attempts/ female/ season			
mean number of successful broods/ female/ season		0.24	Females only raise a single successful brood per year, but Cowardin and Johnson (1979) estimate only 28% of hens hatch a nest (i.e., hen success rate) and 77% of those broods survive to fledging (flight age), so number of successful nests/female is 0.28 * 0.77 = 0.22. Cowardin et al (1985) reported hen success was only 15% in ND study from 1977-80. They also conclude that to be stable, hen success rate would need to be 31%, so 0.31 * 0.77 = 0.24 successful broods/female in stable population
mean number of fledglings/ successful nest	fpsn	5.0	"Broods that survived to fledging averaged 5.0 ducklings" (Orthmeyer and Ball 1990). Cowardin & Johnson (1979) present estimates from several sites in Table 3 ranging from 4.87 to 5.40. Assume mean of 5.0 fledglings/successful nest.
mean number of fledglings/ female/ season (ARS)	ARS	1.19	1.18 fledglings/female from Cowardin and Johnson 1979, Table 4. Based on Cowardin et al. (1985) if hen success = 31%, the 0.31 * 0.77 * 5.0 = 1.19 fledglings/female in stable population.

Reference List

Alisauskas, R. T., and C. D. Ankney. 1992. The cost of egg laying and its relationship to nutrient reserves in waterfowl. In: *Ecology and Management of Breeding Waterfowl.* B. Batt (Ed.), University of Minnesota Press. Pp 30-61.

Ball, I. J., D. S. Gilmer, L. M. Cowardin, and J. H. Reichmann. 1975. Survival of wood duck and mallard broods in north-central Minnesota. *J. Wildl. Manage.* 39, no. 4:776-80.

Bellrose, F. C. 1980. Ducks, geese, and swans of North America. 3rd ed. Stackpole Books, Harrisburg, PA

Coulter, M. W. and W. R. Miller. 1968. Nesting biology of Black Ducks and Mallards in northern New England. *Vt. Fish Game Bull.* 38(2):1-73.

Cowardin, L. M., D. S. Gilmer, and C. W. Shaiffer. 1985. Mallard recruitment in the agricultural environment of North Dakota. *Wildl. Monogr.* 92: 1-37.

Cowardin, L. M., and D. H. Johnson. 1979. Mathematics and mallard management. *J. Wildl. Manage.* 43, no. 1:18-35.

Drilling, N., R. Titman, and F. McKinney. 2002. Mallard (*Anas platyrhynchos*), The Birds of North America Online (A. Poole, Ed.). Ithaca: Cornell Lab of Ornithology; Retrieved from the Birds of North America Online: http://bna.birds.cornell.edu/bnaproxy.birds.cornell.edu/bna/species/658

Dzubin, A. and J. B. Gollop. 1972. Aspects of Mallard breeding ecology in Canadian parkland and grassland. Pages 113-152 *in* Population ecology of migratory birds. U.S. Fish Wildl. Serv. Rep. no. 2.

Figley, W. K. and L. W. VanDruff. 1982. Ecology of urban Mallards. *Wildl. Monogr.* 81:1-39.

Johnson, D. H., D. W. Sparling, and L. M. Cowardin. 1987. A model of the productivity of the mallard duck. *Ecol. Model.* 38: 257-75.

Krapu, G. L. 1981. The role of nutrient reserves on Mallard reproduction. *Auk* 98:29-38

Orthmeyer, D. L., and I.J. Ball. 1990. Survival of mallard broods on Benton Lake National Wildlife Refuge in nothcentral Montana. *J. Wildl. Manage.* 54, no. 1: 62-66.

Palmer, R. S. 1976. Handbook of North American birds, Vol. 2: waterfowl. Yale Univ. Press, New Haven, CT.

Swanson, G. A., T. L. Shaffer, J. F. Wolf, and F. B. Lee. 1986. Renesting characteristics of captive mallards on experimental ponds. *J. Wildl. Manage.* 50, no. 1: 32-38.

Sowls, L. K. 1955. Prairie ducks: a study of their behavior, ecology, and management. Wildl. Manage. Inst., Washington, D.C., and Stackpole Co. Harrisburg, PA.

Blue-winged teal (*Anas discors*) — **Four-letter Alpha Code: BWTE**

Species life-history parameters	Model Code	Typical value	Rationale
daily mortality rate during laying & incubation	m1	0.0476	"Following measures of Blue-winged Teal nest success are Mayfield estimates or were converted to Mayfield estimates (Green 1989) and include only studies of ≥60 nests: Iowa = 38% (Bennett 1938), 7% (Glover 1956), 16% (Burgess et al. 1965), 9.5% (Weller 1979), 14% (Fleskes and Klaas 1991), 14% (LaGrange et al. 1995); S. Dakota = 40% (Duebbert and Lokemoen 1976), 32% (Simpson 1988), 24%, and 29% (Klett et al. 1988); N. Dakota = 43% (Klett and Johnson 1982), 34% (Duebbert et al. 1986), 11% and 17% (2 sites; Klett et al. 1988), 23% (Lokemoen et al. 1990a), 10% (Kantrud 1993), 17%, 14%, 26%, 20%, 13%, and 9% (6 sites; P. R. Garrettson and FCR unpubl.); Saskatchewan = 21% (Stoudt 1971) 15%, 23%, and 16% (3 sites, Greenwood et al. 1995), 10% (McKinnon and Duncan 1999); Alberta = 32% (Keith 1961), 14% (Smith 1971), 27% (Clark et al. 1991), 12% (Greenwood et al. 1995)" as quoted from Rohwer et al 2002. The mean of all of these estimates is 20%. Assuming a 20% nest success rate over a 33 d laying and incubation period, the daily nest mortality rate is 0.0476.
daily mortality rate during nestling-rearing	m2	0.0046	"In Manitoba, 83% of broods fledged ≥1 young (Rohwer 1985)" as quoted from Rohwer et al 2002. Over a 40 day period from hatching to fledging, this translates to a daily brood mortality rate of 0.0046.
date of first egg of first nest (dd-mmm)	T1	May-1	"In nw. Iowa, nesting begins late Apr, with peak of nest-initiation dates in early May; latest nests initiated early Jul (Bennett 1938, Glover 1956, Strohmeyer 1967). In sw. Manitoba, nesting begins early May and peaks mid-May, with latest nest starts in mid-Jul (Sowls 1955, Dane 1965, FCR). In N. Dakota, arrival dates vary with temperature, but typically 12–16 Apr; nesting begins 1–13 May" as quoted from Rohwer et al 2002. Assume typical egg laying period from May 1 to July 10.
date of first egg of last nest (dd-mmm)	Tlast	Jul-10	
length of rapid follicle growth period (RFG) for each egg (days)	rfg	5	Based on allometric equation for waterfowl from Alisauskas and Ankney (1992), assume rfg = 5 d.
mean clutch size	clutch	10	"Data from 45 studies throughout breeding range: mean clutch size for completed nests 10.1 (range 6–14, n = 5,634 nests)" as quoted from Rohwer et al. 2002.
mean intra-egg laying interval (days)	eli	1	"Eggs almost always laid at a rate of 1 egg/d, except in early spring when cold and wet weather can cause skips in daily laying pattern (ERL). Time to complete clutch is usually equal to number of eggs laid" as quoted from Rohwer et al. 2002.
egg on which female typically begins incubation–penultimate (1) or last (0)	penult	0	"Nocturnal incubation typically begins day last egg is laid, although unusually low ambient temperatures may delay onset of nocturnal incubation for up to 5 d (Loos 1999)" as quoted from Rohwer et al. 2002.
duration from start of incubation to hatch (days)	I	23	Duration of incubation period ranges from 19–29 d and significantly decreases over season, with a mean of 22.9 d (Feldheim 1997). Assume typical duration is 23 d.

Species Life-History Profiles – 12 December 2013

Description	Code	Value	Notes
duration from hatch to fledging of nestlings (days)	N	40	"First flight of captive-reared birds at 40 d of age, soft landing at 41 d, and aerial maneuvers at 43 d (Dane 1965). Wild birds fledge at about 40 d (Clark et al. 1988)" as quoted from Rohwer et al. 2002.
duration since nest failure due to other reasons until female initiates new nest (days)	We	6	"Renest intervals, the time between nest destruction and initiation of another clutch, do not appear to depend on date or incubation stage (Strohmeyer 1967, Carlson 1981)" as quoted from Rohwer et al. 2002. Assume renesting occurs rapidly after loss, with typical value 1 d longer than rfg.
duration since successful fledging until female initiates new nest (days)	Wf	100	Only one successful brood per year. Given the length of time needed to raise successful brood to fledging, the egglaying period defined above is not sufficient to allow renesting after success, consequently, the value for Wf will not affect model outcomes, so use default value of 100 d.
female body weight (g) during breeding season	BdyWt	400	Based on data in Table 3 of Rohwer et al. (2002), female weights are approximately 400 g just prior to and during egg laying.
diet composition during breeding season	O		"Primary foods of breeding females: snails (36%), aquatic insects (36%), crustaceans (16%), annelids (Annelida, 2%), and plant material, including seeds (9%; Swanson et al. 1974)" as quoted from Rohwer et al. 2002. T-REX does not include categories for aquatic invertebrates, but assume adult diet consisting of 91% invertebrates and 9% seeds and juvenile diet of 100% invertebrates.
mean number of nest attempts/ female/ season			"Females will renest if first nest destroyed, but renesting rates relatively low, around 34% in studies in Iowa and Manitoba (Strohmeyer 1967, Rohwer 1986a) and only 22% in Wisconsin (Carlson 1981)" as quoted from Rohwer et al. 2002.
mean number of successful broods/ female/ season			
mean number of fledglings/ successful nest	fpsn	5.8	In Manitoba, Rohwer (1985) reported that 29 unmanipulated nests had an average of 10.2 ± 1.01 hatchlings and 5.8 ± 4.6 fledglings. Even though he defined fledging as 20 days old, "no ducklings were lost after 20 days in the 7 broods that were observed repeatedly up to flying age" (Rohwer 1985). Assume that 5.8 fledglings per successful nest is typical.
mean number of fledglings/ female/ season (ARS)	ARS		

Reference List

Alisauskas, R. T., and C. D. Ankney. 1992. The cost of egg laying and its relationship to nutrient reserves in waterfowl. In: *Ecology and Management of Breeding Waterfowl.* B. Batt (Ed.), University of Minnesota Press. Pp 30-61.

Bennett, L. J. 1938. The Blue-winged Teal: its ecology and management. Collegiate Press, Inc. Ames, IA.

Burgess, H. H., H. H. Prince, and D. L. Trauger. 1965. Blue-winged Teal nesting success as related to land use. *J. Wildl. Manage.* 29:89-95.

Carlson, G. R. 1981. Renesting in Blue-winged Teal at Horicon National Wildlife Refuge, Wisconsin. Master's Thesis. Univ. of Wisconsin, Madison.

Clark, R. G., T. D. Nudds, and R. O. Bailey. 1991. Populations and nesting success of upland-nesting ducks in relation to cover establishment. *Can. Wildl. Serv. Progr. Notes* 193.

Clark, R. G., L. G. Sugden, R. K. Brace, and D. J. Nieman. 1988. The relationship between nesting chronology and vulnerability to hunting of dabbling ducks. *Wildfowl* 39:137-144.

Dane, C. W. 1965. The influence of age on the development and reproductive capability of the Blue-winged Teal (*Anas discors* Linnaeus). Phd Thesis. Purdue Univ. Lafayette, IN.

Duebbert, H. F. and J. T. Lokemoen. 1976. Duck nesting in fields of undisturbed grass-legume cover. *J. Wildl. Manage.* 40:39-49.

Duebbert, H. F., J. T. Lokemoen, and D. E. Sharp. 1986. Nest sites of ducks in grazed mixed-grass prairie in North Dakota. *Prairie Nat.* 18:99-108.

Feldheim, C. L. 1997. The length of incubation in relation to nest initiation date and clutch size in dabbling ducks. *Condor* 99:997-1001.

Fleskes, J. P. and E. E. Klaas. 1991. Dabbling duck recruitment in relation to habitat and predators at Union Slough National Wildlife Refuge, Iowa. U.S. Fish Wildl. Serv. Fish Wildl. Tech. Rep. 32.

Glover, F. A. 1956. Nesting and production of the Blue-winged Teal (*Anas discors* Linnaeus) in northwest Iowa. *J. Wildl. Manage.* 20:28-46.

Green, R. E. 1989. Transformation of crude proportions of nests that are successful for comparison with Mayfield estimates of nest success. *Ibis* 131:305-306.

Greenwood, R. J., A. B. Sargeant, D. H. Johnson, L. M. Cowardin, and T. L. Shaffer. 1995. Factors associated with duck nest success in the prairie pothole region of Canada. Wildl. Monogr. no. 128Kantrud, H. A. 1993. Duck nest success on Conservation Reserve Program land in the prairie pothole region. *J. Soil Water Conserv.* 48:238-242.

Keith, L. B. 1961. A study of waterfowl ecology on small impoundments in southeastern Alberta. *Wildl. Monogr.* 6.

Klett, A. T. and D. H. Johnson. 1982. Variability in nest survival rates and implications to nesting studies. *Auk* 99:77-87.

Klett, A. T., T. L. Shaffer, and D. H. Johnson. 1988. Duck nest success in the prairie pothole region. *J. Wildl. Manage.* 52:431-440.

Lagrange, T. G., J. L. Hansen, R. D. Andrews, A. W. Hancock, and J. M. Kienzler. 1995. Electric fence predator exclosure to enhance duck nesting: a long-term case study in Iowa. *Wildl. Soc. Bull.* 23:261-266.

Lokemoen, J. T., H. F. Duebbert, and D. E. Sharp. 1990a. Homing and reproductive habits of Mallards, Gadwalls, and Blue-winged Teal. *Wildl. Monogr.* no. 106.

Loos, E. R. 1999. Incubation in Blue-winged Teal (*Anas discors*): testing hypotheses of incubation constancy, recess frequency, weight loss, and nest success. Master's Thesis. Louisiana State Univ., Baton Rouge.

Mckinnon, D. T. and D. C. Duncan. 1999. Effectiveness of dense nesting cover for increasing duck production in Saskatchewan. *J. Wildl. Manage.* 63:382-389.

Rohwer, F. C. 1985. The adaptive significance of clutch size in prairie ducks. *Auk* 102:354-361.

Rohwer, F. C. 1986a. The adaptive significance of clutch size in waterfowl. Phd Thesis. Univ. of Pennsylvania, Philadelphia.

Rohwer, F.C., W.P. Johnson, and E.R. Loos. 2002. Blue-winged Teal (*Anas discors*), The Birds of North America Online (A. Poole, Ed.). Ithaca: Cornell Lab of Ornithology; Retrieved from the Birds of North America Online: http://bna.birds.cornell.edu/bnaproxy.birds.cornell.edu/bna/species/625

Simpson, S. G. 1988. Duck nest success on South Dakota game production areas. Pages 140-145 *in* Eighth Great Plains Wildlife Damage Control Workshop Proceedings. (Uresk, D. W., G. L. Schenbeck, and R. Cefkin, Eds.) USDA For. Serv. Gen. Tech. Rep. RM-154.

Smith, A. G. 1971. Ecological factors affecting waterfowl reproduction in the Alberta parklands. U.S. Fish Wildl. Serv. Resour. Publ. 98.

Sowls, L. K. 1955. Prairie ducks: A study of their behavior, ecology, and management. Stackpole Co., Harrisburg, PA, and Wildl. Manage. Inst. Washington, D.C.

Stoudt, J. H. 1971. Ecological factors affecting waterfowl production in the Saskatchewan parklands. U.S. Fish Wildl. Serv. Resour. Publ. 99.

Strohmeyer, D. L. 1967. The biology of renesting by the Blue-winged Teal (*Anas discors*) in northwest Iowa. Phd Thesis. Univ. of Minnesota, St. Paul.

Swanson, G. A., M. I. Meyer, and J. R. Serie. 1974. Feeding ecology of breeding Blue-winged Teals. *J. Wildl. Manage.* 38:396-407.

Weller, M. W. 1979. Density and habitat relationships of Blue-winged Teal nesting in northwestern Iowa. *J. Wildl. Manage.* 43:367-374.

Northern bobwhite (*Colinus virginianus*)			Four-letter Alpha Code: NOBO
Species life-history parameters	Model Code	Typical value	Rationale
daily mortality rate during laying & incubation	m1	0.039	Burger et al. (1995) reported daily nest survival rates for female first nest and renests of 0.9692 and 0.9458, respectively. Since 64% of female-incubated nest were first nests, the weighted average nest survival for the whole season is 0.961. Daily nest mortality rate is 0.039.
daily mortality rate during nestling-rearing	m2	0.016	For bobwhite, we define "fledging" as chicks reaching 2 weeks of age, because both parents brood chicks for the first two weeks until capable of partial thermoregulation. After two weeks brooding gradually decreases and chicks are fully capable of thermoregulation at 30 d (Brennan 1999). "Overall brood success rate was 0.80 (defined as >1 chick surviving to 2 weeks of age)" from DeVos and Mueller (1993).
date of first egg of first nest (dd-mmm)	T1	10-May	Default dates based on Figure 4 in Brennan 1999. Reported dates vary considerably by region and weather.
date of first egg of last nest (dd-mmm)	Tlast	15-Sep	
length of rapid follicle growth period (RFG) for each egg (days)	rfg	6	Assume 6 d based on reported RFG for chukar, California quail, and Japanese quail ranging from 5 to 7 d.
mean clutch size	clutch	13	Average 12-14 eggs (range 7-28) (Brennan 1999).
mean intra-egg laying interval (days)	eli	1.4	"Rate of laying is slightly <1/d. Takes 18 d to lay clutch of 12-15 eggs (Rosene 1969: 71)" from Brennan (1999). Assume eli = 1.4, so a clutch of 13 eggs is formed over about 17 d.
egg on which female typically begins incubation—penultimate (1) or last (0)	penult	0	"Incubation begins after the last egg is laid." Quote from Brennan (1999).
duration from start of incubation to hatch (days)	I	23	average 23 d ± 1 d (Stoddard 1931, Rosene 1969).
duration from hatch to fledging of nestlings (days)	N	14	Hatching is synchronous, usually all within 1 d (Stoddard 1931, Rosene 1969). For bobwhite, we define "fledging" as chicks reaching 2 weeks of age, because both parents brood chicks for the first two weeks until chicks are capable of partial thermoregulation and being to fly. After two weeks brooding gradually decreases and chicks are fully capable of thermoregulation at 30 d (Brennan 1999).
duration since nest failure due to other reasons until female initiates new nest (days)	We	9	average 9 d (range 1-45 d) (Burger pers. comm. in Brennan 1999).
duration since successful fledging until female initiates new nest (days)	Wf	100	Young stay with adult into late winter or early spring, so if female is caring for a successful brood, she will not renest that season. Assume Wf of 100 d, but any value for Wf that would end after the last day of egg laying (Tlast) will not affect the model results.

female body weight (g) during breeding season	BdyWt	170	Female average of 170 g (n=692; Nelson and Martin 1953).
diet composition during breeding season		O	Adult females during breeding season 80% seeds, 20% invertebrates. "Diet of chicks consists of >80% insects and arthropods for the first two weeks of life" (quote from Brennan 1999), so assume average chick diet is 90% invertebrates and 10% seeds.
mean number of nest attempts/ female/ season			
mean number of successful broods/ female/ season			
mean number of fledglings/ successful nest	fpsn	4.3	Based on data in DeVos and Mueller (1993): 12.8 eggs/clutch * 0.88 mean hatch rate in successful nests * 0.38 mean survival rate of chicks to 14d = 4.3 fledglings (14 d old) /successful nest.
mean number of fledglings/ female/ season (ARS)	ARS		
Notes: Although in some bobwhite populations a female may lay a clutch of eggs that is incubated by the male while she immediately lays a second clutch that she incubates, the above profile does not incorporate male-incubated clutches, only female-incubated clutches.			

Reference List

Brennan, L.A. 1999. Northern Bobwhite (Colinus virginianus), The Birds of North America Online (A. Poole, Ed.). Ithaca: Cornell Lab of Ornithology; Retrieved from the Birds of North America Online: http://bna.birds.cornell.edu.bnaproxy.birds.cornell.edu/bna/species/397

Burger, L. W. Jr., T. V. Dailey, E. W. Kurzejeski, and M. R. Ryan. 1995. Survival and cause-specific mortality of northern bobwhite in Missouri. *J. Wildl. Manage.* 59, no. 2: 401-10.

DeVos, T., and B. S. Mueller. 1993. Reproductive ecology of northern bobwhite in north Florida. *Quail III: National Quail Symposium.* Church, K. E.//Dailey, T. V., 83-90. Kansas Department of Wildlife and Parks, Pratt, KS.

Rosene, W. 1969. The bobwhite quail: its life and management. Rutgers Univ. Press, New Brunswick, NJ.

Stoddard, H. L. 1931. The bobwhite quail: its life history and management. Charles Scribner's Sons, New York.

American kestrel (*Falco sparverius*)

Four-letter Alpha Code: AMKE

Species life-history parameters	Model Code	Typical value	Rationale
daily mortality rate during laying & incubation	m1	0.0042	In MT & WY from 1977-85, 87% (213/246) of monitored nest produced fledglings (Wheeler 1992). In CA from 1977-80, 82% (53/65) of nests produced fledglings (Bloom and Hawks 1983). In WI from 1968-1972, 81% (51/63) of nests produced fledglings (Hamerstrom et al. 1973). In MO from 1982-1984, 70% (46/66) of nests produced fledglings (Toland and Elder 1987). In IA from 1988-1992, 65% (92/142) of nests produced fledglings (Varland and Loughling 1993). In IA from 1989-1992, 74% (48/65) of nests produced fledglings (Craft and Craft 1996). Weighted mean (503/647) = 77.8%; converted to daily nest mortality rate using 60 day = 0.0042.
daily mortality rate during nestling-rearing	m2	0.0042	
date of first egg of first nest (dd-mmm)	T1	Apr-1	Based on Figure 5 in Smallwood and Bird (2002), set start and end dates in Iowa as April 1 to Jun 9 and in Florida as March 11 to Jun 9.
date of first egg of last nest (dd-mmm)	Tlast	Jun-9	
length of rapid follicle growth period (RFG) for each egg (days)	rfg	7	Assume same as Eurasian kestrel (*Falco tinnunculus*) (Meijer et al. 1989).
mean clutch size	clutch	5	Most often 4 or 5 (Smallwood and Bird 2002). Table 8 lists mean clutch sizes for several sites with an overall mean of 4.6.
mean intra-egg laying interval (days)	eli	2	"Generally, 1 egg laid every other day, occasionally 1 or 3 d apart (Bird and Palmer 1988); captives averaged 2.4 d (Porter and Wiemeyer 1972)." Quote from Smallwood and Bird (2002).
egg on which female typically begins incubation—penultimate (1) or last (0)	penult	1	" Female typically begins incubation upon laying penultimate egg" as quoted from Smallwood and Bird (2002).
duration from start of incubation to hatch (days)	I	30	" Usually 27-29 d in captivity, approximately 30 d in the wild (range 26-32)" as quoted from Smallwood and Bird (2002).
duration from hatch to fledging of nestlings (days)	N	30	"Typically at age 28-31 d." as quoted from Smallwood and Bird (2002).
duration since nest failure due to other reasons until female initiates new nest (days)	We	12	"Kestrel readily lays replacement clutch upon loss of first, generally 11-12 d later (Bowman and Bird 1985)" as quoted from Smallwood and Bird (2002).
duration since successful fledging until female initiates new nest (days)	Wf	14	"Dependent on parents for food for about 12-14 d after fledging (Bird and Palmer 1988, Varland et al. 1991)" as quoted from Smallwood and Bird (2002).
female body weight (g) during breeding season	BdyWt	120	Females 120 g ± 9.2 SD (n=111; Dunning 1993). From Smallwood and Bird (2002).

Species Life-History Profiles – 12 December 2013

diet composition during breeding season	I	Although kestrels feed primarily in invertebrates and small mammals, for this species profiles we assume both adults and juveniles consume 100% invertebrates. Current OPP exposure models provide no estimate for residue concentrations in vertebrate food items.
mean number of nest attempts/ female/ season		
mean number of successful broods/ female/ season		Based on observation that "approximately 11% of pairs (n=325) initiated second clutch after successful first attempt" (Smallwood and Bird 2002).
mean number of fledglings/ successful nest	fpsn 4.0	In CA from 1977-80 mean of 3.7 fledglings/successful nest (112/30 nests) (Bloom and Hawks 1983). In MT and WY, from 1977-85 mean of 4.2 fledglings/successful nest (899/213 nests) (Wheeler1992). In IA Varland and Loughin (1993) report 2.9 fledglings per brood, but if calculated on basis of successful broods = 3.75 fledglings/successful brood (345/92). In IA from 1989-92 mean of 4.0 fledglings/successful nest (132/33) (Craft and Craft 1996). Weighted mean = 4.04 (1488/368).
mean number of fledglings/ female/ season (ARS)	ARS	

Reference list

Bird, D. M. and R. S. Palmer. 1988. American Kestrel. Pages 253-290 in Handbook of North American birds. Vol. 5: diurnal raptors. Pt. 2. (Palmer, R. S., Ed.) Yale Univ. Press, New Haven, CT.

Bloom, P. H. and S. J. Hawks. 1983. Nest box use and reproductive biology of the American Kestrel in Lassen County, California. Raptor Res. 17:9-14.

Bowman, R. and D. M. Bird. 1985. Reproductive performance of American Kestrels laying replacement clutches. Can. J. Zool. 63:2590-2593.

Dunning, Jr., J. B. 1993. CRC handbook of avian body masses. CRC Press, Boca Raton, FL.

Hamerstrom, F., F. N. Hamerstrom, and J. Hart. 1973. Nest boxes: an effective management tool for kestrels. J. Wildl. Manage. 37:400-403.

Porter, R. D. and S. N. Wiemeyer. 1972. Reproductive patterns in captive American Kestrels (Sparrow Hawks). Condor 74:46-53.

Smallwood, J.A. and D. M. Bird. 2002. American Kestrel (Falco sparverius), The Birds of North America Online (A. Poole, Ed.). Ithaca: Cornell Lab of Ornithology; Retrieved from the Birds of North America Online: http://bna.birds.cornell.edu/bnaproxy.birds.cornell.edu/bna/species/602

Toland, B. R. and W. H. Elder. 1987. Influence of nest-box placement and density on abundance and productivity of American Kestrels in central Missouri. Wilson Bull. 99:712-717.

Varland, D. E. and T. M. Loughin. 1993. Reproductive success of American Kestrels nesting along an interstate highway in central Iowa. *Wilson Bull.* 105:465-474.

Wheeler, A. H. 1992. Reproductive parameters for free ranging American Kestrels (*Falco sparverius*) using nest boxes in Montana and Wyoming. *J. Raptor Res.* 26:6-9.

Killdeer (*Charadrius vociferus*) | **Four-letter Alpha Code: KILL**

Species life-history parameters	Model Code	Typical value	Rationale
daily mortality rate during laying & incubation	m1	0.03	" Kantrud and Higgins (1992) found that young were produced at 39% of nests ($n = 79$) in native grassland and 57% of nests ($n = 27$) in cropland in n-central U.S. and s-central Canada. In Ontario, Nol (1980), using Mayfield (1975) method, found success rates of 28.9% for 17 nests on Long Point, 64% for 12 nests on adjacent mainland. Of 101 nests observed in Mississippi over 4 yr, 38 (37.6%) hatched at least 1 egg (Schardien 1981)" as quoted from Jackson and Jackson (2000). In MN, Mace (1971) found 12 or 26 nests (46%) hatched at least one egg. Weighted mean of percent nest success = 41.6%, which over a 29 d laying and incubation period translates to a daily nest failure rate = 0.03.
daily mortality rate during nestling-rearing	m2	0.048	No specific data for directly parameterizing m2. Following broods after hatching is complicated by the fact that some broods are split between the male and female and sometimes juveniles may join other broods, so determining if a brood successfully fledges juveniles is very difficult. Schardien (1981) reported that 38 successfully hatched nests produced 132 hatchlings of which 16 fledged. Mace (1971) reported 12 successful nests produced 23 hatchlings of which 7 fledged. Combined: 23 fledglings from 50 nests that hatched chicks. Although no information was reported on the mean number of fledglings from successfully fledged nests, the number must be greater than 1 and less than 4 (typical # of eggs). If we assume that the typical number of fledglings per successful nest is 2, then we can estimate that 23% of nests successfully fledged juveniles (i.e., (23/2)/50). Over a 30 d period from hatching to fledging, this translates to a daily brood mortality rate of 0.048. Assume 0.048 is a typical value for m2.
date of first egg of first nest (dd-mmm)	T1	Mar-15	"First clutches laid in early Mar in Mississippi (Schardien 1981); mid-Mar in N. Carolina (Jensen 1948) and Maryland (Stewart and Robbins 1958); late Mar in Pennsylvania (Miller 1949); mid- to late Mar in ne. California (L. W. Oring pers. comm.); and mid-Apr in Washington State (Jewett et al. 1953) and Minnesota (Mace 1971). In n. Michigan, egg-laying begins in late Apr and continues through Jun; most first nest attempts initiated during first half of May (Powell and Cuthbert 1993). In Minnesota, up to 3 clutches laid per season, but no pair raised >1 brood; active nests found mid-Apr–late Jul (Mace 1971). In s. U.S., successful hatching of 2 broods/season may be common; ≥3 are possible (Schardien 1981)" as quoted from Jackson and Jackson (2000). Based on Figure 5 of Jackson and Jackson (2000), assume typical egg laying dates from March 15 to July 15.
date of first egg of last nest (dd-mmm)	Tlast	Jul-15	
length of rapid follicle growth period (RFG) for each egg (days)	rfg	5	Based on Roudybush et al. (1979), RFG period in related species: Ruddy turnstone (*Arenaria interpres*), 5-6 d; Western sandpiper (*Calidris mauri*), 5-8 d; Red phalarope (*Phalaropus fulicarius*), 4-5 d; and Northern phalarope (*Phalaropus lobatus*), 6-7 d. Assume typical RFG period in killdeer is 5 d.
mean clutch size	clutch	4	"Typically 4 eggs; reports of fewer may reflect lost eggs. Reports of 5 are rare and usually poorly documented (Townsend 1929, Miller 1933, Stone 1937, Nickell 1943)" as quoted from Jackson and Jackson (2000).

mean intra-egg laying interval (days)	eli	1	"Eggs usually laid in morning at 1-d, much less often 2-d, intervals (Furniss 1933)" as quoted from Jackson and Jackson (2000).
egg on which female typically begins incubation—penultimate (1) or last (0)	penult	0	"In Mississippi, steady incubation begins with laying of last (usually fourth) egg (JAJ, BJSJ)" as quoted from Jackson and Jackson (2000).
duration from start of incubation to hatch (days)	I	25	"Varies considerably; probably associated with ambient temperatures, frequency of disturbance at nest, and possibly human variation in measuring. In Mississippi, 22–28 d (mean 25.1 ± 1.6 SD, n = 16 nests; Schardien 1981); in Michigan, 24.7–25.9 d (mean 25.1, n = 4 nests; Bunni 1959); at 1 Tennessee nest, 29.5 d (Ganier 1934). Many reports of incubation periods between 23 and 29 d are reviewed in Schardien 1981 and Bunni 1959" as quoted from Jackson and Jackson (2000). Assume typical duration of 25 d.
duration from hatch to fledging of nestlings (days)	N	30	"Ability to fly is attained by 20 (Demaree 1975) to 31 (Lenington 1980) d of age. Schardien (1981) found that some chicks in Mississippi attained ability to fly by age 30 d, whereas others did not. Chicks remain in close association with parents and siblings until they can fly" as quoted from Jackson and Jackson (2000). Assume chicks become independent and are considered "fledged" at 30 d of age.
duration since nest failure due to other reasons until female initiates new nest (days)	We	6	"The shortest period of time elapsing between loss of either eggs or chicks and completion of the new clutch was 8 days, meaning the nest was probably initiated about 3 days after the previous nest was lost" as quoted from Schardien (1981). Given the assumption of a 5-d RFG period for killdeer, assume the typical duration of We period is 6 d.
duration since successful fledging until female initiates new nest (days)	Wf	6	"Schardien (1981) observed marked chicks with their parents regularly until age 39 d, and 1 instance of a chick foraging with its parents at age 81 d. In most northern areas, second or third nesting attempts are result of failure of earlier efforts, although second brood is occasionally hatched after successful first brood (e.g., Michigan; Bunni 1959, Nickell 1943)" as quoted from Jackson and Jackson (2000). Schardien (1981) reported the duration from hatching the first nest attempt until the completion of the second clutch for four nest attempts that fledged chicks (i.e., 38, 57, 39, and 39 d). Assuming 4 d to complete a clutch and 30 d to raise hatchlings to fledging, the duration of Wf for there four nests would be 4, 23, 5, and 5 d, for a mean of 9 d. Assume renesting after fledging typically occurs quickly and a typical value for Wf is 6 d.
female body weight (g) during breeding season	BdyWt	101	Female mean = 101 g, range = 87.7-121, n=6 and male mean = 92.1 ± 10.4, range = 83.9-109, n=10 (Dunning 1984).
diet composition during breeding season	I		Both adults and juveniles feed primarily on terrestrial invertebrates, though some adults consume small amounts of plant material, primarily seeds (Jackson and Jackson 2000). Assume both adults and juveniles consume 100% invertebrates.
mean number of nest attempts/ female/ season			In MS, Schardien (1981) estimated an average of 1.94 nest attempts per pair.
mean number of successful broods/ female/ season			

	fpsn	
mean number of fledglings/ successful nest	2	No specific data on the number of fledglings per successful nest. See the rationale described for m2 above. The number of fledglings must be greater than 1 and less than 4 (typical # of eggs). Assume a typical value is 2 fledglings per successfully fledged nest attempt.
mean number of fledglings/ female/ season (ARS)	ARS 0.8	"In Minnesota, Lenington (1980) reported overall annual success of 1.6 independent young, and Mace (1971) an average of 0.5 young per pair" as quoted from Jackson and Jackson (2000). In MS, Schardien (1981) estimates an average of 0.4 fledglings per pair per year. The mean of these three studies is 0.8 fledglings per female per year.

Reference List

Bunni, M. K. 1959. The Killdeer (*Charadrius v. vociferus*), Linnaeus, in the breeding season: ecology, behavior, and the development of homoiothermism. Ph.D. thesis. Univ. of Michigan, Ann Arbor.

Demaree, S. R. 1975. Observations on roofnesting Killdeers. *Condor* 77:487-488.

Furniss, O. C. 1933. Observations on the nesting of the Killdeer Plover in the Prince Albert district in central Saskatchewan. *Can. Field-Nat.* 47:135-138.

Ganier, A. F. 1934. Incubation period of the Killdeer. *Wilson Bull.* 46:17-19.

Jackson, Bette J. and Jerome A. Jackson. 2000. Killdeer (Charadrius vociferus), The Birds of North America Online (A. Poole, Ed.). Ithaca: Cornell Lab of Ornithology; Retrieved from the Birds of North America Online: http://bna.birds.cornell.edu.bnaproxy.birds.cornell.edu/bna/species/517

Jensen, O. F. 1948. Killdeers nest early. *Chat* 12:53.

Jewett, S. A., W. P. Taylor, W. T. Shaw, and J. W. Aldrich. 1953. Birds of Washington state. Univ. of Washington Press, Seattle.

Lenington, S. 1980. Bi-parental care in Killdeer: an adaptive hypothesis. *Wilson Bull.* 92:8-20.

Kantrud, H. A. and K. F. Higgins. 1992. Nest and nest site characteristics of some ground-nesting, non-passerine birds of northern grasslands. *Prairie Nat.* 24:67-84.

Mace, T. R. 1971. Nest dispersion and productivity of Killdeers, *Charadrius vociferus*. Master's Thesis. Univ. of Minnesota, St. Paul.

Mayfield, H. 1975. Suggestions for calculating nest success. *Wilson Bull.* 87:456-466.

Miller, R. F. 1933. The breeding birds of Philadelphia, P.A. Oologist 50:86-95Miller, R. F. 1949. The Killdeer. Breeding birds of the Philadelphia region (Part IV). *Cassinia* 37:1-8.

Nickell, W. P. 1943. Observations on the nesting of the Killdeer. *Wilson Bull.* 55:23-28.

Nol, E. 1980. Factors affecting the nesting success of the Killdeer (*Charadrius vociferus*) on Long Point, Ontario. Master's Thesis. Univ. of Guelph, Guelph, ON.

Powell, A. N. and F. J. Cuthbert. 1993. Augmenting small populations of plovers: an assessment of cross-fostering and captive-rearing. *Conserv. Biol.* 7:160-168.

Roudybush, T. E., C. R. Grau, M. R. Petersen, D. G. Ainley, K. V. Hirsch, A. P. Gilman, and S. M. Patten. 1979. Yolk formation in some charadriiform birds. *The Condor* 81: 293-98.

Schardien, B. J. 1981. Behavioral ecology of a southern population of Killdeer. Phd Thesis. Mississippi State Univ. Mississippi State.

Stewart, R. E. and C. S. Robbins. 1958. Birds of Maryland and the District of Columbia. *N. Am. Fauna* no. 62.

Stone, W. 1937. Bird studies at old Cape May. Delaware Valley Ornithol. Club, Philadelphia, PA.

Townsend, C. W. 1929. *Oxyechus vociferus* (Linnaeus) Killdeer. Pages 202-217 *in* Life histories of North American shorebirds. Pt. 2. (Bent, A. C., Ed.) U.S. Natl. Mus. Bull. 146.

White-winged dove (*Zenaida asiatica*)			Four-letter Alpha Code: WWDO
Species life-history parameters	Model Code	Typical value	Rationale
daily mortality rate during laying & incubation	m1	0.020	"Nest-survival probability (probability that a nest, once established, will produce fledglings) for urban-nesting doves in San Antonio, TX, calculated using Mayfield (1961) method, averaged 0.456 ± 0.402–0.510 95% CI ($n = 397$) over 2 yr (West et al. 1993). Hayslette and Hayslette (1999) reported nest survival of 0.575 ± 0.068 SE ($n = 57$) in a Kingsville, TX, urban colony. All losses occurred prior to hatching" as quoted from Schwertner et al. (2002). Also in TX, Small et al. (2005) calculated that the Mayfield nest success rate for all nests for both years was 0.518 (SE = 0.006; n=34). The weighted mean nest success rate from the 3 studies is 0.52. Over a 32 d nest period (i.e., 15+16+2-1), the daily nest mortality rate is 0.020.
daily mortality rate during nestling-rearing	m2	0.020	(same studies) the daily nest mortality rate is 0.020.
date of first egg of first nest (dd-mmm)	T1	15-May	"In Texas, breeds from late Apr to late Aug (Passmore 1981). Usually 2 nesting peaks/season; timing varies annually (Cunningham et al. 1997)" as quoted from Schwertner et al. (2002). "The data obtained in the Santa Cruz thicket definitely proves two rather distinct periods of nesting, and indicates the rearing of two broods by a consider-able proportion of the population. The first-brood nests are built in mid-May, and the young birds leave during the latter part of June. The second-brood nests are begun late in June and early in July, and the young are fledged in late July and early August" as quoted from Neff (1940). Based on the observations of Neff (1940) and Figure 3 of Schwertner et al. (2002), assume that the core period of egg laying starts May 15 and ends July 15.
date of first egg of last nest (dd-mmm)	Tlast	15-Jul	
length of rapid follicle growth period (RFG) for each egg (days)	rfg	5	Based on allometric equation (RFG = 2.852*Egg mass ^0.31) from Alisauskas and Ankney (1992), using an egg mass of 7.7 g, assume rfg = 5 d.
mean clutch size	clutch	2	"Generally 2 eggs. In lower Rio Grande valley of Texas, 43 of 987 (4.4%) active nests contained 3 eggs (Cottam and Trefethen 1968). Near Phoenix, AZ, 384 nests (91%) contained 2 eggs, 38 nests (9%) 1 egg, 1 nest (<1%) 3 eggs (Neff 1940)" as quoted from Schwertner et al. (2002).
mean intra-egg laying interval (days)	eli	1	"Eggs usually laid 24 h apart" as quoted from Schwertner et al. (2002).
egg on which female typically begins incubation–penultimate (1) or last (0)	penult	0	"Incubation generally starts after laying of second egg" as quoted from Schwertner et al. (2002).
duration from start of incubation to hatch (days)	I	16	"Varies individually and geographically. In s. Texas, 14 d with <1 d variation. In Arizona, usually 15–17 d, although occasionally up to 20 d (Neff 1940)" as quoted from Schwertner et al. (2002). Assume the typical incubation is 16 d.
duration from hatch to fledging of nestlings (days)	N	15	"Usually fledge at 13–18 d of age, although capable at 8 d (Neff 1940, Cottam and Trefethen 1968)" as quoted from Schwertner et al. (2002). Assume the typical nestling period is 15 d.
duration since nest failure due to other reasons until female initiates new nest (days)	We	7	"Usually renests immediately following predation of previous nest" as quoted from Schwertner et al. (2002). No specific data found on the duration of We, but assume renesting after failure occurs rapidly and with a 5-d rfg period, the typical period for We is 7 d.

Species Life-History Profiles – 12 December 2013

duration since successful fledging until female initiates new nest (days)	Wf	10	"Subsequent nesting initiated 4–15 d following fledging of previous brood (Neff 1940, Saunders 1940, Alamia 1970, Blankinship 1970)" as quoted from Schwertner et al. (2002). Assume a typical period for Wf is 10 d.
female body weight (g) during breeding season	BdyWt	153	Dunning (1993) reported a mean mass of 153 g ± 13.2 SD ($n = 30$) for Arizona birds" as quoted from Schwertner et al. (2002).
diet composition during breeding season	G		"Primarily seeds, mast, and fruit (Dolton 1975, Haughey 1986, Cunningham et al. 1997). Grains and other agricultural crops, where available, dominate diet across much of range (Dolton 1975, West 1993, Cunningham et al. 1997). Both adults feed milky secretion formed in crop glands ("crop milk"), as in other columbids. First-day feedings may contain fruit pulp and soft seeds. By 3–4 d, feedings contain foods more similar to adult diet" as quoted from Schwertner et al. (2002). Assume both adults and juveniles consume 100% seeds.
mean number of nest attempts/ female/ season			
mean number of successful broods/ female/ season			"Early authors contended second broods were common (Saunders 1940, Cottam and Trefethen 1968). Alamia (1970) and Williams (1971), however, reported relatively low rates of renesting (25% and 9%, respectively) by color-tagged birds, although dense vegetation may have made resighting marked birds difficult (Swanson 1989). Schacht et al. (1995) used radiotelemetry to monitor White-winged Doves in lower Rio Grande valley. Of 39 birds that successfully hatched their first brood, 27 (69%) attempted a second nest. Success rate of second attempt not reported" as quoted from Schwertner et al. (2002).
mean number of fledglings/ successful nest	fpsn	1.8	No specific data found on the number of fledglings/successful nest. Based on data from Neff (1940), since 384 nests (91%) contained 2 eggs, 38 nests (9%) 1 egg, 1 nest (<1%) 3 eggs, the mean number of eggs/nest is 1.9, and if probability of fledging is not affected by clutch size, the maximum number of fledglings/successful nest would also be 1.9. Hayslette et al. (2000) reported that 93.5% of hatchlings survive to fledging. Until empirical data is found, assume that most successful nests experience little loss of eggs or nestlings and that the typical number of fledglings/successful nest is 1.8.
mean number of fledglings/ female/ season (ARS)	ARS		"Cottam and Trefethen (1968) reported seasonal production of 1.9 young/breeding pair in s. Texas. They estimated total number of breeding pairs as number of active nests at peak of nesting season. In Arizona, 1955–1959, seasonal production was 0.95–1.64 (mean 1.44) juveniles/adult, based on ratio of juveniles to adults in harvest (Stair 1970)" as quoted from Schwertner et al. (2002).

Reference list

Alamia, L. A. 1970. Renesting activity and breeding biology of the White-winged Dove (*Zenaida asiatica*) in the lower Rio Grande valley of Texas. Master's Thesis. Texas A&M Univ., College Station.

Alisauskas, R. T., and C. D. Ankney. 1992. The cost of egg laying and its relationship to nutrient reserves in waterfowl. In: *Ecology and Management of Breeding Waterfowl*. B. Batt (Ed.), University of Minnesota Press. Pp 30-61.

Blankinship, D. R. 1970. White-winged Dove nesting colonies in northeastern Mexico. *Trans. N. Am. Wildl. Nat. Resour. Conf.* 35:171-182.

Cottam, C. and J. B. Trefethen. 1968. White-wings: the life history, status, and management of the White-winged Dove. D. Van Nostrand Co., Inc. Princeton, NJ.

Cunningham, S. C., R. W. Engel-Wilson, P. M. Smith, and W. B. Ballard. 1997. Food habits and nesting characteristics of sympatric Mourning and White-winged doves in Buckeye-Arlington Valley, Arizona. Tech. Rep. no. 26. Arizona Game Fish Dep. Phoenix.

Dolton, D. D. 1975. Patterns and influencing factors of White-winged Dove feeding activity in the lower Rio Grande valley of Texas and Mexico. Master's Thesis. Texas A&M Univ., College Station.

Dunning, Jr., J. B. 1993. CRC handbook of avian body masses. CRC Press, Inc., Boca Raton, FL.

Haughey, R. A. 1986. Diet of desert-nesting western White-winged Doves *Zenaida asiatica mearnsi*. Master's Thesis. Univ. of Arizona, Tucson.

Hayslette, S. E. and B. A. Hayslette. 1999. Late and early season reproduction of urban White-winged Doves in southern Texas. *Tex. J. Sci.* 51:173-180.

Hayslette, S. E., T. C. Tacha, and G. L. Waggerman. 2000. Factors affecting white-winged, white-tipped, and mourning dove reproduction in lower Rio Grande valley. *J. Wildl. Manage.* 64:286-295.

Mayfield, H. F. 1961. Nesting success calculated from exposure. *Wilson Bull.* 73:255-261.

Neff, J. A. 1940. Notes on nesting and other habits of the Western White-winged Dove in Arizona. *J. Wildl. Manage.* 4:279-290.

Passmore, M. F. 1981. Population biology of the Common Ground Dove and ecological relationships with Mourning and White-winged doves in south Texas. Phd Thesis. Texas A&M Univ., College Station.

Saunders, G. B. 1940. Eastern White-winged Dove (*Melopelia asiatica asiatica*) in southeastern Texas. U.S. Fish Wild. Ser. Rep.

Schacht, S. J., T. C. Tacha, and G. L. Waggerman. 1995. Bioenergetics of White-winged Dove reproduction in the lower Rio Grande valley of Texas. *Wildl. Monogr.* 129:1-31.

Schwertner, T. W., H. A. Mathewson, J. A. Roberson, M. Small and G. L. Waggerman. 2002. White-winged Dove (Zenaida asiatica), The Birds of North America Online (A. Poole, Ed.). Ithaca: Cornell Lab of Ornithology; Retrieved from the Birds of North America Online: http://bna.birds.cornell.edu/bna/species/710.

Small, M. F., C. L. Schaefer, J. T. Bacchus and J. A. Roberson. 2005. Breeding ecology of white-winged doves in a recently colonized urban environment. *Wilson Bull.* 117:172-176.

Stair, J. L. 1970. Chronology of the nesting season of White-winged Doves *Zenaida asiatica mearnsi* (Ridgway) in Arizona. Master's Thesis. Univ. of Arizona, Tucson.

Swanson, D. A. 1989. Breeding biology of the White-winged Dove (*Zenaida asiatica*) in south Texas. Master's Thesis. Texas A&I Univ., Kingsville.

West, L. M. 1993. Ecology of breeding White-winged Dove in the San Antonio metropolitan area. Master's Thesis. Texas Tech Univ., Lubbock.

West, L. M., L. M. Smith, R. S. Lutz, and R. R. George. 1993. Ecology of urban White-winged Doves. *Trans. N. Am. Wildl. Nat. Resour. Conf.* 59:70-77.

Williams, L. 1971. A renesting study of the White-winged Dove (*Zenaida asiatica asiatica* [Linnaeus]) in the lower Rio Grande valley of Texas. Master's Thesis. Texas A&I Univ., Kingsville.

Mourning dove (*Zenaida macroura*)	Four-letter Alpha Code: MODO		
Species life-history parameters	Model Code	Typical value	Rationale
daily mortality rate during laying & incubation	m1	0.025	Average 48% nest success (Sayre and Silvy 1993) over 29 d nest period = 0.975 daily success rate; average 47% nest success in Iowa (Iowa DNR website: http://www.iowadnr.gov/Portals/idnr/uploads/education/Species/birds/mdove.pdf)
daily mortality rate during nestling-rearing	m2	0.025	
date of first egg of first nest (dd-mmm)	T1	15-Mar	First brood may occur any time during year along Gulf Coast, but generally occurs late Feb-early March at southern latitudes and 1-2 mo later at more northern latitudes (Mirarchi and Baskett 1994); assume typical values for T1 of 15 March (julian date 75) and Tlast of August 30 (i.e., 245-75=170) (Mirarchi and Baskett 1994).
date of first egg of last nest (dd-mmm)	Tlast	30-Aug	
length of rapid follicle growth period (RFG) for each egg (days)	rfg	6	Based on 5-7 d for ring dove and 5-8 d for domestic pigeon (King 1973).
mean clutch size	clutch	2	(Mirarchi and Baskett 1994).
mean intra-egg laying interval (days)	eli	2	Second egg laid on consecutive days or alternate days (24-48 hrs, most common) (Mirarchi and Baskett 1994).
egg on which female typically begins incubation–penultimate (1) or last (0)	penult	0	Incubation starts with laying of second egg.
duration from start of incubation to hatch (days)	I	14	(Mirarchi and Baskett 1994).
duration from hatch to fledging of nestlings (days)	N	15	Normally 15 d but can be earlier if frightened (Mirarchi and Baskett 1994).
duration since nest failure due to other reasons until female initiates new nest (days)	We	6	"After a nesting failure, the period until a new clutch is begun ranges from 2 to 25 days, with the most frequent time interval being 6 days in one study (Hanson and Kossack 1963) and 3-5 days in another (Swank 1955).
duration since successful fledging until female initiates new nest (days)	Wf	4	"Interval between completion of 1 clutch and initiation of next depends on the fate of the first clutch (Hanson and Kossack 1963). Approximately 30 d are necessary between initiation of successful clutch and initiation of subsequent clutch. Eggs may be laid in same or another nest while young from previous nest are still being tended" (Mirarchi and Baskett 1994); Mourning doves usually begin a new clutch within two to five days after the young fledge (Blockstein and Westmoreland 1993).
female body weight (g) during breeding season	BdyWt	116	female *Z. m. carolinensis* – 123 g and *Z. m. marginella* 108 (from Mirarchi and Baskett 1994); female: 115 g ±1.76 (n=95)(Dunning 1984).

	G		
diet composition during breeding season			Adults consume primarily seeds, with invertebrates and green forage taken only incidentally (Mirarchi and Baskett 1994). Juveniles initially fed crop milk produced by parents but gradually fed more regurgitated seeds after 3-4 d of age until diet similar to adults at fledging. Assume both adults and juveniles consume 100% seeds.
mean number of nest attempts/ female/ season			5-6 clutches per year in southern states and fewer in north (Mirarchi and Baskett 1994); 3 to 5 nests per year (Iowa DNR website).
mean number of successful broods/ female/ season			
mean number of fledglings/ successful nest	fpsn	1.85	1.8-1.9 young per successful nest (Iowa DNR web site).
mean number of fledglings/ female/ season (ARS)	ARS	3.6	3.6 young fledged per breeding pair per year from Sayre and Silvy 1993 in Mirarchi and Baskett 1994.

Reference List

Blockstein, D. E., and D. Westmoreland. 1993. Reproductive strategy. In *Ecology and Management of the Mourning Dove*. T.S. Baskett, M.W. Sayre, R.E. Tomlinson, and R.E. Mirarchi (Eds.), pp. 105-16. Harrisburg, PA: Stackpole Books.

Hanson, H. C. and C. W. Kossack. 1963. The Mourning Dove in Illinois. Illinois Dept. Conserv. Tech. Bull. 2, Southern Illinois Univ. Press, Carbondale.

King, J. R. 1973. Energetics of reproduction in birds. Breeding Biology of Birds. Pp. 78-107. Washington, DC: National Academy of Sciences.

Mirarchi, R. E. 1993. Care and propagation of captive mourning doves. In *Ecology and Management of the Mourning Dove*. T. S. Baskett, M.W. Sayre, R.E. Tomlinson, and R.E. Mirarchi, (Eds.), pp. 409-28. Harrisburg, PA: Stackpole Books.

Mirarchi. R. E. and T. S. Baskett. 1994. Mourning dove (Zenaida macroura). In The Birds of North America, No. 117 (A. Poole and F. Gill, Eds.) The Academy of Natural Sciences, Philadelphia and The American Ornithologists' Union, Washington, DC.

Sayre, M. W. and N. J. Silvy. 1993. Nesting and Production. In *Ecology and Management of the Mourning Dove*. T.S. Baskett, M.W. Sayre, R.E. Tomlinson, and R.E. Mirarchi (Eds.), pp. 81-104. Harrisburg, PA: Stackpole Books.

Swank, W. B. 1955. Feather molt as an aging technique for mourning doves. *J Wildl. Manage.* 19: 412-414.

Westmoreland, D. and L. B. Best. 1985. The effect of disturbance on mourning dove nesting success. *The Auk* 102: 774-80.

———. 1987. What limits mourning doves to a clutch of two eggs? *The Condor* 89: 486-93.

Westmoreland, D., L. B. Best, and D. E. Blockstein. 1986. Multiple brooding as a reproductive strategy: Time-conserving adaptations in mourning doves. *The Auk* 103: 196-203.

Northern flicker (*Colaptes auratus*)

Four-letter Alpha Code: NOFL

Species life-history parameters	Model Code	Typical value	Rationale
daily mortality rate during laying & incubation	m1	0.022	"Of 119 nests where eggs were laid, 93 (78%) survived to hatching" (Moore 1995); $0.978^{11} = 0.78$; $1 - 0.978 = 0.022$.
daily mortality rate during nestling-rearing	m2	0.012	80 of 93 nests (86%) survived to an advanced hatchling stage (all nests were monitored to the stage when hatchlings could be banded, usually 11-14 d after hatching, and many were monitored until near fledging). (Moore 1995): $0.988^{13} = 0.86$; $1 - 0.988 = 0.012$.
date of first egg of first nest (dd-mmm)	T1	7-May	From Moore (1995) – In NE modal date of first egg between 9 and 18 May (n=30, range 1 May –1 June); in WY modal dates between 7 and 12 May (n=17, range April 29 – 17 June); in NM modal date between 1 and 6 May (n=17, range April 27-June 5);
date of first egg of last nest (dd-mmm)	Tlast	20-Jun	From Ingold (1996) – 36 pairs in Ohio laid from first week of May until 4th week of June.
length of rapid follicle growth period (RFG) for each egg (days)	rfg	5	No specific information found on flickers or other woodpeckers. Based on allometric equation (RFG = $2.852 * Egg\ mass^{0.31}$) from Alisauskas and Ankney (1992), using and egg mass of 7.01 g, assume rfg = 5 d.
mean clutch size	clutch	7	Clutch size pooled for all latitudes: mean= 6.5 (SD 1.4; range 3-12, n=411) (Moore and Koenig 1986).
mean intra-egg laying interval (days)	eli	1	Eggs laid at rate of 1/day, usually between 0500 and 0600 (Sherman 1910).
egg on which female typically begins incubation–penultimate (1) or last (0)	penult	1	Incubation begins 1 or 2 days before last egg is laid (Sherman 1910).
duration from start of incubation to hatch (days)	I	11	Sherman's (1910) observations: mean = 11 d (SD 1.06, n=8).
duration from hatch to fledging of nestlings (days)	N	25	"For yellow-shafted flicker in n. Florida, F. Lohrer (unpubl. data) recorded a mean nestling period of 24 d (range 21-26); from Sherman (1910) for Iowa, 25 d (range 24-27)" (Moore 1995).
duration since nest failure due to other reasons until female initiates new nest (days)	We	14	Although flickers sometime use old nests, assume that renesting, especially after starling competition or depredation, occurs in new nest. Mean excavation time for new nest is 12.1 d in Ontario (range 5-19, n=10) (Lawrence 1966) and 15.2 d in WI (range 11-20, n=15) (Burkett 1989), so excavation time may determine fastest renesting time.
duration since successful fledging until female initiates new nest (days)	Wf	21	"Duration of juvenile dependence on parents appears to vary but is not long." (Moore 1995).
female body weight (g) during breeding season	BdyWt	130	Mean mass (g ± SD, range, N): *Yellow-shafted flicker* – NE 131.83 ± 7.46, 117-150, n=20 (Short 1965); PA 129.0 ± 7.67, 106-164, n=65 (Dunning 1993); *Red-shafted flicker* – CO 139.17 ± 4.63,130-145, n=6 (Short 1965); OR 142.0 ± 2.2, 121-167, n=82 (Dunning 1993); *Gilded flicker* – 111.0 ± 7.36, 92-129, n=100 (Dunning 1993).

Species Life-History Profiles – 12 December 2013

Parameter	Code	Value	Description
diet composition during breeding season		O	Based on annual analysis by Beal (1911, as cited in Moore 1995), yellow-shafted flickers (n=684) consume 60.9% invertebrates and 39.1% plants (fruit and seeds) while red-shafted flickers consume 67.7% invertebrates and 32.3 % plants. About 50% of the total diet consists of ants. Most of the plant material seems to be fruits. However, during the breeding season (i.e., May and June), figures in Martin et al (1951) indicate that adults consume about 90% invertebrates and 10% fruits. Juveniles fed by regurgitation. Assume both adults and juveniles consume 90% invertebrates and 10% fruit.
mean number of nest attempts/ female/ season			
mean number of successful broods/ female/ season			"Apparently single brooded. Neither yellow-shafted nor red shafted flickers regularly produce a second brood in a breeding season, but it may happen occasionally." (Moore 1995); only 1 of 31 flicker pairs successfully raised two broods (Ingold 1996); although no information on how many of the females with successful broods attempted a second clutch, the probability of quitting after success is apparently high.
mean number of fledglings/ successful nest	fpsn	6.15	The mean number of 10-17 d old chicks from 79 nests at 4 locations is 6.15 (Moore and Koenig 1986).
mean number of fledglings/ female/ season (ARS)	ARS		

References

Alisauskas, R. T., and C. D. Ankney. 1992. The cost of egg laying and its relationship to nutrient reserves in waterfowl. In: Ecology and Management of Breeding Waterfowl. B. Batt (Ed.), University of Minnesota Press. Pp 30-61.

Beal, F. E. 1911. Food of the woodpeckers of the United States. U. S. Dept. Agric. Biol. Surv. Bull. 37.

Burkett, E. W. 1989. Differential roles of sexes in breeding northern flickers (*Colaptes auratus*). Ph.D. diss., Univ. of Wisconsin, Milwaukee.

Dunning, J. B. 1993. CRC handbook of avian body masses. CRC Press, Boca Raton, FL.

Ingold, D. 1994. Nest-site characteristics of the red-bellied and red-headed woodpeckers and northern flickers in east-central Ohio. *Ohio J. Sci.* 94: 2-7.

Lawrence, L. D. 1966. A comparative life-history study of four species of woodpeckers. *Ornithol. Monogr. 5.*

Martin, A. C., H. S. Zim, and A. L. Nelson. 1951. *American wildlife and plants: A guide to wildlife food habits.* Dover Publications, Inc., New York. 500 pp.

Moore, W. S. 1995. Northern flicker (*Colaptes auratus*). In The Birds of North America, No. 166 (A. Poole and F. Gill, Eds.). The Academy of Natural Sciences, Philadelphia, and the American Ornithologists' Union, Washington, DC.

Moore, W. S. and W. D. Koenig. 1986. Comparative reproductive success of yellow-shafted, red-shafted, and hybrid flickers across a hybrid zone. *The Auk* 103:42-51.

Sherman, A. 1910. At the sign of the northern flicker. *Wilson Bull.* 22:135-171.

Short, L. L. 1965. Hybridization in the flickers (*Colaptes*) of North America. *Bull. Am. Mus. Nat. Hist.* 129:307-428.

Willow flycatcher (*Empidonax traillii*)			Four-letter Alpha Code: WIFL
Species life-history parameters	Model Code	Typical value	Rationale
daily mortality rate during laying & incubation	m1	0.0187	"Nest success (nests producing ≥1 fledgling) variable. In Wisconsin, 315 of 459 nests (68.6%) were successful (McCabe 1991); the Mayfield method (Mayfield 1961, 1975) yielded more modest nest success of 51% (McCabe 1991). In British Columbia, nest success 28% (*n* = 96 nests; Campbell et al. 1997); much higher in s. Michigan, where 65.2% of 92 nests fledged young (Walkinshaw 1966). In Ohio and Nebraska, Holcomb (1972b) reported 39.5% nest success (*n* = 91 nests). In sw. New Mexico (1997–1999), 43.3% of 298 nests fledged ≥1 young (S. Stoleson and D. Finch pers. comm.), and in Sierra Nevada, 60% (*n* = 25) and 50% (*n* = 64) of nests fledged ≥1 young in 1997 and 1998, respectively (H. Bombay pers. comm.)." as quoted from Sedgwick (2000). The weighted mean of above nest success rates is 54.6%, which over a 32 d nesting period (i.e., 14+14+4) results in a daily nest mortality rate of 0.0187.
daily mortality rate during nestling-rearing	m2	0.0187	
date of first egg of first nest (dd-mmm)	T1	12-Jun	"Earliest and latest dates for full clutches in Washington: 25 May and 13 Jul (Jewett et al. 1953), and 19 Jun and 24 Jul (King 1955). In Wisconsin, McCabe (1991) reported mean first egg as 27 Jun. Mean first egg date in s. Michigan 17 Jun (earliest 11 Jun) (Walkinshaw 1966). In s. Sierra Nevada, earliest and latest egg dates 25 May and 29 Jul (M. Whitfield pers. comm.). For first nests in se. Oregon, mean first egg date 22 Jun (median 22 Jun, range 5–30 Jun, *n* = 439 nests)" as quoted from Sedgwick (2000). The mean of the start and end dates listed above is June 12 and July 22, respectively. This is similar to the core egg-laying dates in Figure 3 of Sedgwick (2000).
date of first egg of last nest (dd-mmm)	Tlast	22-Jul	
length of rapid follicle growth period (RFG) for each egg (days)	rfg	3	Based on allometric equation (RFG = 2.852*Egg mass ^0.31) from Alisauskas and Ankney (1992), using an egg mass of 1.7 g, assume rfg = 3 d.
mean clutch size	clutch	4	"Mean of 3.68 ± 0.1 SE for first nests (*n* = 31) and 3.14 ± 0.1 SE for renests (*n* = 29) in Nebraska (Holcomb 1974). In Wisconsin, McCabe (1991) reported a mean clutch size of 3.59 ± 0.49 SD (*n* = 415 clutches), with 58% of clutches being 4 eggs; early (before 28 Jun) clutches (mean 3.68, *n* = 243) larger (*p* <0.001) than late clutches (mean 3.49, *n* = 172), and first clutches (mean 3.5) larger than renest clutches (mean 3.2) for the same pairs (*n* = 21). In se. Oregon (1988–1997), mean first nest (unparasitized) clutch size 3.69 ± 0.03 SE (range 1–5, *n* = 365 clutches); 69.6% were 4 eggs and 26.9% were 3 eggs (JAS). In s. California (*E. t. extimus*) first, second, and third clutch sizes: 3.63 ± 0.05 SE (*n* = 113), 2.90 ± 0.09 SE (*n* = 50) and 2.71 ± 0.19 SE (*n* = 14), respectively (M. Whitfield pers. comm.). In s. New Mexico, mean clutch size reported as 3.06 ± 0.63 SE (*n* = 50; includes second and later nestings; S. Stoleson pers. comm.). *E. t. extimus* first nests in Arizona (1996–1999, unparasitized nests only) had average clutch of 2.92 ± 0.73 SD (*n* = 321; T. McCarthey pers. comm.)." as quoted from Sedgwick (2000). Assume typical clutch size of 4 eggs.
mean intra-egg laying interval (days)	eli	1.3	"One egg/d; often 1 day is skipped, so 4-egg clutch complete in 5 d" as quoted from Sedgwick (2000).

Description	Code	Value	Notes
egg on which female typically begins incubation—penultimate (1) or last (0)	penult	1	"Unknown at what stage of egg-laying female typically begins sitting on eggs continuously at night. Given that eggs hatch over a 1- to 3-d interval (McCabe 1991, JAS), eggs must be maintained for periods long enough for embryonic growth prior to laying of last egg, perhaps similar to incubation behavior in Dusky Flycatcher (Morton and Pereyra 1985; Sedgwick 1993a, 1993b)" as quoted from Sedgwick (2000).
duration from start of incubation to hatch (days)	I	14	"In Wisconsin, 14.8 d ($n = 50$; McCabe 1991); in Nebraska and Ohio, 13.3 d $\pm$ 0.1 SE (range 12–14, $n = 28$ nests; Holcomb 1972a); in s. Michigan, incubation periods were 15 d at 3 nests, 14 d at 3 nests, and 13 d at 1 nest (from last egg laid to last or all eggs hatched; Walkinshaw 1966)" as quoted from Sedgwick (2000).
duration from hatch to fledging of nestlings (days)	N	14	"Nestling period 14–15 d (Berger and Hofslund 1950, McCabe 1991). Berger (1967) reported 13–16 d for 45 young. Five family groups in s. Michigan fledged after an average of 13.8 d in nest ($n = 13$ young; Walkinshaw 1966); in a Nebraska and Ohio study, 82 young fledged between 11 and 14 d (mean 12.3 d $\pm$ 0.1 SE; Holcomb 1972a)" as quoted from Sedgwick (2000).
duration since nest failure due to other reasons until female initiates new nest (days)	We	7	"Mean of 6.6 d ($n = 18$ renests; Holcomb 1974) and 6.5 d ($n = 21$ renests; McCabe 1991) between first nest loss and initial egg-laying in renest" as quoted from Sedgwick (2000).
duration since successful fledging until female initiates new nest (days)	Wf	100	"Normally only 1 brood/season except in cases of predation or nest loss. Renesting after successfully fledging a brood is rare in northern populations (1 instance, $n = 882$ pairs, $n = 1,168$ nests; 1988–1997; se. Oregon; JAS), somewhat more common farther south (M. Whitfield pers. comm.)" as quoted from Sedgwick (2000). Since females typically raise only a single successful brood per year, the value for Wf needs to be long enough so renesting after success does not occur – set default to 100 d.
female body weight (g) during breeding season	BdyWt	12.46	"In Michigan, average mass of breeding males, 12.9 g (range = 11.4–14.7, $n = 18$); breeding females, 12.3 g (range 10.2–14.2, $n = 22$; Walkinshaw 1966). Mean mass of breeding season adults from se. Oregon: males 12.72 g $\pm$ 0.70 SD ($n = 373$); females 12.47 g $\pm$ 1.12 SD ($n = 369$; Sedgwick and Klus 1997)" as quoted from Sedgwick (2000). Weighted mean = 12.46 g.
diet composition during breeding season		I	Both adults and juveniles consume diet consisting of almost entirely invertebrates.
mean number of nest attempts/ female/ season			
mean number of successful broods/ female/ season			
mean number of fledglings/ female/ successful nest	fpsn	3.14	Based on data from Sedgwick (2000) in WI and MI, the weighted mean of number of fledglings/successful nest is 3.14.
mean number of fledglings/ female/ season (ARS)	ARS	1.81	In OR over 10 yrs: Seasonal fecundity: mean of 1.81 $\pm$ 0.05 SE young fledged/female ($n = 874$ females; Sedgwick and Iko 1999).

Reference list

Alisauskas, R. T., and C. D. Ankney. 1992. The cost of egg laying and its relationship to nutrient reserves in waterfowl. In: Ecology and Management of Breeding Waterfowl. B. Batt (Ed.), University of Minnesota Press. Pp 30-61.

Berger, A. J. 1967. Traill's Flycatcher in Washtenaw County, Michigan. *Jack-Pine Warbler* 45:117-123

Berger, A. J. and P. B. Hofsland. 1950. Notes on the nesting of the Alder Flycatcher (*Empidonax traillii*) at Ann Arbor, Michigan. *Jack-Pine Warbler* 28:7-11.

Campbell, R. W., N. K. Dawe, I. McTaggart-Cowan, J. M. Cooper, and G. W. Kaiser. 1997. The birds of British Columbia, Vol. 3-Passerines: flycatchers through vireos. R. Br. Columbia Mus. Victoria.

Holcomb, L. C. 1972a. Traill's Flycatcher breeding biology. *Nebr. Bird Rev.* 40:50-68.

Holcomb, L. C. 1972b. Nest success and age-specific mortality in Traill's Flycatchers. *Auk* 89:837-841.

Holcomb, L. C. 1974. The influence of nest building and egg laying behavior on clutch size in renests of the Willow Flycatcher. *Bird-Banding* 45:320-325.

Jewett, S. G., W. P. Taylor, W. T. Shaw, and J. W. Aldrich. 1953. Birds of Washington State. Univ. of Washington Press, Seattle.

King, J. R. 1955. Notes on the life history of Traill's Flycatcher (*Empidonax traillii*) in southeastern Washington. *Auk* 72:148-173.

Mayfield, H. 1961. Nesting success calculated from exposure. *Wilson Bull.* 73:255-261.

Mayfield, H. 1975. Suggestions for calculating nest success. *Wilson Bull.* 87:456-466.

McCabe, R. A. 1991. The little green bird. Ecology of the Willow Flycatcher. Rusty Rock Press, Madison, WI.

Morton, M. L. and M. E. Pereyra. 1985. The regulation of egg temperatures and attentiveness patterns in the Dusky Flycatcher (*Empidonax oberholseri*). *Auk* 102:25-37.

Sedgwick, J. A. 1993a. Reproductive ecology of the Dusky Flycatcher in western Montana. *Wilson Bull.* 105:84-92.

Sedgwick, J. A. 1993b. Dusky Flycatcher (*Empidonax oberholseri*). *in* The birds of North America, no. 78. (Poole, A., P. Stettenheim, and F. Gill, Eds.) Acad. Nat. Sci., Philadelphia, PA, and Am. Ornithol. Union, Washington, DC.

Sedgwick, J. A. 2000. Willow Flycatcher (*Empidonax traillii*), The Birds of North America Online (A. Poole, Ed.). Ithaca: Cornell Lab of Ornithology; Retrieved from the Birds of North America Online: http://bna.birds.cornell.edu.bnaproxy.birds.cornell.edu/bna/species/533

Sedgwick, J. A. and W. M. Iko. 1999. Costs of Brown-headed Cowbird parasitism to Willow Flycatchers. *Stud. Avian Biol.* 18:167-181.

Sedgwick, J. A. and R. J. Klus. 1997. Injury due to leg bands in Willow Flycatchers. *J. Field Ornithol.* 68:62-629.

Walkinshaw, L. H. 1966. Summer biology of Traill's Flycatcher. *Wilson Bull.* 78:31-46.

Eastern phoebe (*Sayornis phoebe*)			Four-letter Alpha Code: EAPH
Species life-history parameters	Model Code	Typical value	Rationale
daily mortality rate during laying & incubation	m1	0.0097	Daily nest mortality rate of 0.0097 based on 37 d period (Weeks1979).
daily mortality rate during nestling-rearing	m2	0.0097	
date of first egg of first nest (dd-mmm)	T1	Apr-10	In southern IN, two peaks in laying (mid April and late May/early June) with eggs starting about April 10 and last nest starting about June 21 (Weeks 1979). Similar egg laying dates reported in KS and IL (Klaas 1970, Graber et al. 1974, Weeks 1994). In northern WI, also two peaks in laying (early June and mid July) with first eggs starting May 7 and ending by August 10 (Faanes 1980). Base typical dates on data from IN.
date of first egg of last nest (dd-mmm)	Tlast	Jun-21	
length of rapid follicle growth period (RFG) for each egg (days)	rfg	4	Based on allometric equation (RFG = 2.852*Egg mass ^0.31) from Alisauskas and Ankney (1992), using and egg mass of 2.09 g, assume rfg = 4 d.
mean clutch size	clutch	5	Varies from 2-6 eggs, with mode of 5. Weeks (1994) summarizes clutch sizes from studies in several locations with means ranging from 4.4 to 4.8 eggs/clutch.
mean intra-egg laying interval (days)	eli	1	Lay 1 egg/d (Weeks 1979).
egg on which female typically begins incubation—penultimate (1) or last (0)	penult	0	Begin diurnal incubation with the last egg (Weeks 1994).
duration from start of incubation to hatch (days)	I	16	Mean of 16 d (Weeks 1994).
duration from hatch to fledging of nestlings (days)	N	16	Young depart nest at about 16 d (Weeks 1994),
duration since nest failure due to other reasons until female initiates new nest (days)	We	10	Interval of 9-10 d needed to begin egg production after nest loss (Klaas 1970 as reported in Weeks 1978). Klaas observed that this interval was a day or longer than interval after successful nest because females can initiate physiological changes prior to fledging. Assume 10 d is a typical interval.
duration since successful fledging until female initiates new nest (days)	Wf	9	Both parents regularly fed fledglings up to 2 weeks post-fledging, but young were rebuffed on 18 d post-fledging (Weeks 1994). In IN, renesting interval after success was 7.8 and 13.3 in 1970 and 1971, respectively (Weeks 1978). Klaas (1970) reported interval of 7.5 days in KS. Based on mean of these two sites, assume 9 d period is typical.
female body weight (g) during breeding season	BdyWt	16.9	Mean 16.9 g, std dev 0.62, range 16.0 to 17.8, n=10 (Weeks 1994).

diet composition during breeding season		I	While phoebes consume small amounts of small fruit during fall, winter, and early spring, diet consists primarily of flying insects. Assume during breeding season than both adults and juveniles consume 100% insects.
mean number of nest attempts/ female/ season			
mean number of successful broods/ female/ season			
mean number of fledglings/ successful nest	fpsn	4.0	In IN, 4.32 (n=71) and 4.07 (n=55) fledglings/successful nest in 1970 & 1971, respectively (Weeks 1979). In KS, 3.8 fledglings/successful nest during 1962-1965 (Klaas 1970) and 1980-1983 (Murphy 1994). In WI, 4.13 (n=29), 3.85 (n=27), and 3.87 (n=32) fledglings/successful nest in 1974, 1975, and 1976, respectively (Faanes 1980). Based on data from all three states, assume mean of 4.0 fledglings/successful nest.
mean number of fledglings/ female/ season (ARS)	ARS	5.8	In IN, 6.76 and 5.14 fledglings/female/ year in 1970 & 1971, respectively (Weeks 1979). In 1992, 5.48 fledglings/female/year (Weeks 1994). Mean of 3 years is 5.8 fledglings/female/year.

Reference list

Alisauskas, R. T., and C. D. Ankney. 1992. The cost of egg laying and its relationship to nutrient reserves in waterfowl. In: Ecology and Management of Breeding Waterfowl. B. Batt (Ed.), University of Minnesota Press. Pp 30-61.

Faanes, C. A. 1980. Breeding biology of Eastern Phoebes in northern Wisconsin. *Wilson Bull*. 92:107-110.

Graber, R. R., J. W. Graber, and E. L. Kirk. 1974. Illinois birds: Tyrannidae. Ill. Nat. Hist. Surv. Biol. Notes No. 86.

Klaas, E. E. 1970. A population study of the Eastern Phoebe and its social relationships with the Brown-headed Cowbird. Phd Thesis. Univ. Kansas, Lawrence.

Murphy, M. T. 1994. Breeding patterns of Eastern Phoebes in Kansas: Adaptive strategies or physiological constraint? *Auk* 111(3):617-633.

Weeks, Jr., H. P. 1978. Clutch size variation in the Eastern Phoebe in southern Indiana. *Auk* 95:656-666.

Weeks, Jr., H. P. 1979. Nesting ecology of the Eastern Phoebe in southern Indiana. *Wilson Bull.* 91:441-454.

Weeks Jr., H. P. 1994. Eastern Phoebe (Sayornis phoebe), In: The Birds of North America, No. 94 (A. Poole and F. Gill, Eds.). Philadelphia: The Academy of Natural Sciences: Washington, D C: The American Ornithologists' Union.

Ash-throated Flycatcher (*Myiarchus cinerascens*)   Four-letter Alpha Code: ATFL

Species life-history parameters	Model Code	Typical value	Rationale
daily mortality rate during laying & incubation	m1	0.0058	"Secondary cavity nester. Nests primarily in natural cavities, woodpecker holes, nest boxes, and cavities in other human-made structures. In U.S. survey, some or all young fledged in >70% of attempts (Phillips 2000). During 2-yr study in s. Washington, 92% of 31 nests (all in natural cavities) successfully fledged young (Seavey 1997, 2000). Ash-throateds breeding in nest boxes during study in lower Colorado River valley were 100% successful (fledged ≥1 young; Brush 1981)" as quoted from Cardiff and Dittmann (2002). Figure 1 of Phillips (2000) shows that 72% of nest attempts fledged young, but includes about 12 of nest attempts where no eggs were laid. If nest success based on nests that received eggs, the apparent nest success in Phillips (2000) is approximately 82%. If overall nest success is assumed to be 82%, over a 34 d nest period (i.e., 15+15+4) the daily nest mortality rate is 0.0058.
daily mortality rate during nestling-rearing	m2	0.0058	
date of first egg of first nest (dd-mmm)	T1	Apr-15	"Reported mid-Mar to late Jun (n = 43; Van Tyne and Sutton 1937, Sutton 1967, Simpkin and Gubanich 1991, Russell and Monson 1998, CNRP), but few well-documented records earlier than mid- to late Apr. Peak ranges from mid- to late Apr in southern lowlands (e.g., LCRV; Brush 1981, Rosenberg et al. 1991) to mid-May at higher elevations in south (e.g., San Jacinto Mtns., CA; Mock et al. 1991) to mid- to late Jun in extreme north (e.g., 14–25 Jun in s. Washington [n = 31]; Seavey 2000)" as quoted from Cardiff and Dittmann (2002). Based on Figure 3 of Cardiff and Dittmann (2002), core nest initiation period from April 15 to June 15. Phillips (2000) reports earliest and latest first egg dates of April 21 to June 11.
date of first egg of last nest (dd-mmm)	Tlast	Jun-15	
length of rapid follicle growth period (RFG) for each egg (days)	rfg	4	Based on allometric equation (RFG = 2.852*Egg mass ^0.31) from Alisauskas and Ankney (1992), using an egg mass of 3.11 g, assume rfg = 4 d.
mean clutch size	clutch	4	"Among 309 egg sets (CM, SBCM, WFVZ), mean clutch size 4.3 (range 2–7); about 44% of sets had 4 eggs (n = 136), 39% had 5 (n = 120), 11% had 3 (n = 33), 4% had 6 (n = 13), 2% had 2 (n = 6), and <1% had 7 (n = 1)" as quoted from Cardiff and Dittmann (2002). Assume typical clutch size of 4.
mean intra-egg laying interval (days)	eli	1	"One egg/day on consecutive days (Bendire 1892, CNRP, J. Seavey pers. comm.)" as quoted from Cardiff and Dittmann (2002).
egg on which female typically begins incubation–penultimate (1) or last (0)	penult	0	"Anecdotal evidence suggests that incubation begins immediately upon completion of clutch (e.g., synchronous hatchings, CNRP)" as quoted from Cardiff and Dittmann (2002).
duration from start of incubation to hatch (days)	I	15	"About 14–16 d (Bendire 1892; Seavey 1997, 2000; CNRP)" as quoted from Cardiff and Dittmann (2002).
duration from hatch to fledging of nestlings (days)	N	15	"Young depart nest at 13–17 d of age (Bendire 1892, Bent 1942, Lanyon 1961, Kaufman 1996, Seavey 2000, CNRP)" as quoted from Cardiff and Dittmann (2002).

duration since nest failure due to other reasons until female initiates new nest (days)	We	6	No specific data found on duration of We. Assume renesting after failure occurs rapidly with a typical duration for We of 6 d.
duration since successful fledging until female initiates new nest (days)	Wf	100	No specific data found on the duration of Wf. "Apparently, no published data establishing that particular pairs of flycatchers raise 2 successive broods" as quoted from Cardiff and Dittmann (2002). Since females typically raise only a single successful brood per year, the value for Wf needs to be long enough so renesting after success does not occur – set default to 100 d.
female body weight (g) during breeding season	BdyWt	33.3	"Means of seasonal subsamples of *M. c. cinerascens* show little deviation from mean of pooled data for each sex; e.g., breeding-season males 28.5 (range 21.2–37.2, $n = 79$), winter males 28.1 (24.8–31.2, $n = 26$), migrant males 29.3 (25.8–33.9, $n =33$), breeding females 27.2 (22–37.8, $n = 52$), winter females 27.4 (25.5–30, $n = 10$), migrant females 26.9 (22.6–35.4, $n = 22$; FMNH, LSUMZ, MVZ, SBCM, UCF). Laying females with enlarged reproductive tracts average heavier (30.9–37.8, mean 33.3, $n = 6$) than pooled sample of breeding-season females (LSUMZ, MVZ, UCF)" as quoted from Cardiff and Dittmann (2002).
diet composition during breeding season		I	During breeding season, almost exclusively arthropods, primarily adult and larval insects.
mean number of nest attempts/ female/ season			
mean number of successful broods/ female/ season			
mean number of fledglings/ successful nest	fpsn	3.6	No specific data found on number of fledglings/successful nest. "In U.S. survey, average number fledglings/nest 3.6 ($n = 65$ nests in nest boxes), and some or all young fledged in >70% of attempts (Phillips 2000)" as quoted from Cardiff and Dittmann (2002). Table 1 of Phillips (2000) indicates that the number of fledglings/nest "includes nests where no young fledged." So, it seems likely that the estimate of 3.6 fledglings/nest underestimates the number of fledglings/successful nests, but by an unknown amount.
mean number of fledglings/ female/ season (ARS)	ARS		

Reference list

Alisauskas, R. T., and C. D. Ankney. 1992. The cost of egg laying and its relationship to nutrient reserves in waterfowl. In: Ecology and Management of Breeding Waterfowl. B. Batt (Ed.), University of Minnesota Press. Pp 30-61.

Bendire, C. 1892. Life histories of North American birds with special reference to their breeding habits and eggs. *U.S. Natl. Mus. Spec. Bull.* 1:263-265.

Bent, A. C. 1942. Life histories of North American flycatchers, larks, swallows, and their allies. *U.S. Natl. Mus. Bull.* 179.

Brush, T. 1981. Response of secondary cavity-nesting birds to manipulation of nest-site availability. Master's Thesis. Arizona State Univ., Tempe.

Cardiff, S. W. and D. L. Dittmann. 2002. Ash-throated Flycatcher (*Myiarchus cinerascens*), The Birds of North America Online (A. Poole, Ed.). Ithaca: Cornell Lab of Ornithology; Retrieved from the Birds of North America Online: http://bna.birds.cornell.edu.bnaproxy.birds.cornell.edu/bna/species/664.

Kaufman, K. 1996. Lives of North American birds. Houghton Mifflin Co. Boston, MA.

Lanyon, W. E. 1961. Specific limits and distribution of Ash-throated and Nutting flycatchers. *Condor* 63:421-449.

Mock, P. J., M. Khubesrian, and D. M. Larcheveque. 1991. Energetics of growth and maturation in sympatric passerines that fledge at different ages. *Auk* 108:34-41.

Phillips, T. 2000. A summary of 1999 data from the birdhouse network. *Birdscope* 14:3.

Rosenberg, K. V., R. D. Ohmart, W. C. Hunter, and B. W. Anderson. 1991. Birds of the lower Colorado River valley. Univ. of Arizona Press, Tucson.

Russell, S. M. and G. Monson. 1998. The birds of Sonora. Univ. of Arizona Press, Tucson.

Seavey, J. 1997. Nest site selection and nesting success of the Ash-throated Flycatcher in south-central Washington. Master's Thesis. Univ. of Washington, Seattle.

Seavey, J. 2000. Nesting success of the Ash-throated Flycatcher in Washington. *Wash. Birds* 7:38-44.

Simpkin, J. L. and A. A. Gubanich. 1991. Ash-throated Flycatchers (*Myiarchus cinerascens*) raise Mountain Bluebird (*Sialia currucoides*) young. *Condor* 93:461-462.

Sutton, G. M. 1967. Oklahoma birds: their ecology and distribution, with comments on the avifauna of the southern Great Plains. Univ. of Oklahoma Press, Norman.

Van Tyne, J. and G. M. Sutton. 1937. The birds of Brewster County, Texas. *Misc. Publ. Univ. of Michigan Mus. Zool.* no. 37:1-119.

Eastern kingbird (*Tyrannus tyrannus*)

Four-letter Alpha Code: EAKI

Species life-history parameters	Model Code	Typical value	Rationale
daily mortality rate during laying & incubation	m1	0.0235	"Nest success (% nests to fledge ≥1 young) over 3-yr period in Kansas varied from 21 to 37% (Murphy 1986a). Over 6 yr in New York, success varied from 28 to 68% (MTM). Nest success in Ontario is comparable (lakeshore = 38%, upland = 52%; Blancher and Robertson 1985a). Based on 13 yr/site records (Kansas, New York, Ontario), 44.5% ± 15.4 SD of kingbird nests fledge young" as quoted from Murphy (1996). For a 34 d nesting period (i.e., 17+15+3-1), an apparent nest success of 44.5% translates to a daily nest mortality rate of 0.0235.
daily mortality rate during nestling-rearing	m2	0.0235	
date of first egg of first nest (dd-mmm)	T1	Jun-1	"Laying dates for first clutches vary across geographic range. In Florida, as early as first or second week of May (Sprunt 1954). In Kansas (Murphy 1983b), New York (Murphy 1983b), and Ontario (Blancher and Robertson 1985a, 1985b), laying generally peaks by first week of Jun. In Manitoba, most first clutches by about second week of Jun (MacKenzie and Sealy 1981)" as quoted from Murphy (1996). Based on Figure 4 in Murphy (1996), core egg laying dates from June 1 to June 30.
date of first egg of last nest (dd-mmm)	Tlast	Jun-30	
length of rapid follicle growth period (RFG) for each egg (days)	rfg	4	Based on allometric equation from Alisauskas and Ankney (1992), assume rfg = 4 d based on 4 g egg.
mean clutch size	clutch	3	"From 2 to 5 eggs; 3 is overall mode; clutches of 4 more common farther north. Average varies geographically from 3.1 ± 0.63 SD (*n* = 42) in w. New York (Murphy 1983b) to 3.7 ± 0.56 (54) in se. Ontario (Blancher and Robertson 1985a) and 3.7 ± 0.75 (135) in British Columbia (M. Funk pers. comm.). Davis (1955) and Murphy (1986a) report mean of 3.4 from Montana (SD 0.50, *n* = 30) and e. Kansas (SD 0.63, *n* = 214), respectively; whereas means of 3.2 ± 0.53 (*n* = 277) and 3.3 ± 0.714 (*n* = 365) recorded in central New York (MTM) and Ontario (Peck and James 1987), respectively" as quoted from Murphy (1996).
mean intra-egg laying interval (days)	eli	1	"One egg laid/d until clutch is complete, but occasionally a day is skipped between laying of penultimate and final egg in clutches of 3 and 4 (Murphy 1983b)" as quoted from Murphy (1996).
egg on which female typically begins incubation—penultimate (1) or last (0)	penult	1	"Females may begin sitting on nest before eggs appear (MTM), but incubation usually begins with laying of penultimate egg in clutches of 3–5" as quoted from Murphy (1996).
duration from start of incubation to hatch (days)	I	15	"Incubation requires 14–17 d, with modes of 15 in New York (mean 15.4 d ± 0.75 SD, *n* = 21) and 14 d in Kansas (14.2 d ± 0.39, *n* = 12; Murphy 1983b)" as quoted from Murphy (1996).
duration from hatch to fledging of nestlings (days)	N	17	"Young leave nest normally 16–17 d after hatching, but ability to maintain strong, level flight does not appear for several more days" as quoted from Murphy (1996).
duration since nest failure due to other reasons until female initiates new nest (days)	We	8	"Pairs in Ontario laid replacement clutches 8 d after initial nest failed (mean 7.7 d, SE 1.7, *n* = 15; Blancher and Robertson 1982b), suggesting that replacement nests require less time for construction" as quoted from Murphy (1996).

duration since successful fledging until female initiates new nest (days)	Wf	35	"Young remain completely dependent on parents for food for at least 2 wk after fledging. Entire family unit normally remains together. Maximum feeding rate during fledgling period is nearly twice that of peak rate during nestling period (Morehouse and Brewer 1968). Young may begin to take fruit and pick small invertebrates off leaves by 25 d, but continue to be fed at high rate until about 5 wk old. Parental feeding continues, but gradually declines for about 2 more weeks (Morehouse and Brewer 1968). Parents defend young against predators up to 7–8 wk of age" as quoted from Murphy (1996).
female body weight (g) during breeding season	BdyWt	40	In KS, male = 42.8 ± 3.76 (39.2–54.7; 24) and females = 41.6 ± 2.67 (37.0–45.8; 32). In NY, males = 38.0 ± 2.36 (33.3–47.0; 92) and females = 37.3 ± 2.08 (33.0–44.7; 105) (from Appendix in Murphy 1996). Assume 40 g is a typical female mass.
diet composition during breeding season			Details from stomach analyses by Beal (1912; $n = 665$) and Dick and Rising (1965; $n = 26$). Beal's analysis conducted over entire breeding range, May–Sep: 85.5% insects; also fruits and seeds of 40 plants. Kingbirds usually take large adult insects, and switch to small prey only when feeding young nestlings or when insect abundance is low (Leck 1971, MTM). Large, whole prey also fed to nestlings" as quoted from Murphy (1996).
mean number of nest attempts/ female/ season			"In New York about 70% of failed first nesting attempts are replaced (mean 68.4%, $n = 57$ nests; MTM). Renesting less common in British Columbia (7 of 36; Siderius 1993). All initial nests that are destroyed while being built are replaced, compared to 82.5% ($n = 40$) of nests lost during incubation, and 35.2% ($n = 17$) that fail during nestling period (MTM). Female may lay up to 3 clutches in 1 season if her nests fail (MTM)" as quoted from Murphy (1996).
mean number of successful broods/ female/ season			"Females raise 1 brood/yr. Taking renesting into account, annual breeding success (percent females to fledge at least 1 nestling) over 6-yr period in New York averaged $58.5\% \pm 15.0$ SD, range 32.9–77.0 (MTM)" as quoted from Murphy (1996).
mean number of fledglings/ successful nest	fpsn	2.5	"Fledged young/successful nest was similar across years and sites: w. New York (2.3) and e. Kansas (2.4; Murphy 1983b), se. Ontario (2.5, 2.7, 3.2; Blancher and Robertson 1985a), and central New York (2.5, 2.5, 2.5, 2.5, 2.7, 2.8; MTM)" as quoted from Murphy (1996). The mean of 4 locations is 2.5 fledglings/successful nest.
mean number of fledglings/ female/ season (ARS)	ARS	1.5	"Average number of fledged young/female over same period was 1.5 nestlings/yr ± 0.40 SD ($n = 6$ yr), range 0.8–1.9" as quoted from Murphy (1996).

References Listed

Alisauskas, R. T., and C. D. Ankney. 1992. The cost of egg laying and its relationship to nutrient reserves in waterfowl. In: *Ecology and Management of Breeding Waterfowl.* B. Batt (Ed.), University of Minnesota Press. Pp 30-61.

Beal, F. E. L. 1912. Food of our more important flycatchers. *U.S. Dep. Agric. Biol. Surv. Bull.* no. 44.

Blancher, P. J. and R. J. Robertson. 1982b. A double-brooded Eastern Kingbird. *Wilson Bull.* 94:212-213.

Blancher, P. J. and R. J. Robertson. 1985a. A comparison of Eastern Kingbird breeding biology in lakeshore and upland habitats. *Can. J. Zool.* 63:2305-2312.

Blancher, P. J. and R. J. Robertson. 1985b. Site consistency in kingbird breeding performance: implications for site fidelity. *J. Anim. Ecol.* 54:1017-1027.

Davis, D. E. 1955. Observations on the breeding biology of kingbirds. *Condor* 57:208-212.

Dick, J. A. and J. D. Rising. 1965. A comparison of foods eaten by Eastern Kingbirds and Western Kingbirds in Kansas. *Bull. Kansas Ornithol. Soc.* 16:23-24.

Leck, C. F. 1971. Some spatial and temporal dimensions of kingbird foraging flights. *Wilson Bull.* 83:310-311.

Mackenzie, D. I. and S. G. Sealy. 1981. Nest site selection in Eastern and Western Kingbirds: a multivariate approach. *Condor* 83:310-312.

Morehouse, E. L. and R. Brewer. 1968. Feeding of nestling and fledgling Eastern Kingbirds. *Auk* 85:44-54.

Murphy, M. T. 1983b. Ecological aspects of the reproductive biology of Eastern Kingbirds: geographic comparisons. *Ecology* 64:914-928.

Murphy, M. T. 1986a. Temporal components of reproductive variability in Eastern Kingbirds (*Tyrannus tyrannus*). *Ecology* 67:1483-1492.

Murphy, M.T. 1996. Eastern Kingbird (*Tyrannus tyrannus*), The Birds of North America Online (A. Poole, Ed.). Ithaca: Cornell Lab of Ornithology; Retrieved from the Birds of North America Online: http://bna.birds.cornell.edu.bnaproxy.birds.cornell.edu/bna/species/253

Peck, G. K. and R. D. James. 1987. Breeding birds of Ontario: nidiology and distribution. Vol. 2. Passerines. R. Ont. Mus. Life Sci. Misc. Publ. Toronto.

Siderius, J. A. 1993. Nest defense in relation to nesting stage and response of parents to repeated model presentations in the Eastern Kingbird (*Tyrannus tyrannus*). *Auk* 110:921-923.

Sprunt, Jr., A. 1954. Florida bird life. Coward-McCann, New York.

Blue jay (*Cyanocitta cristata*)	Four-letter Alpha Code: BLJA		
Species life-history parameters	Model Code	Typical value	Rationale
daily mortality rate during laying & incubation	m1	0.025	Percent nest success varies considerably by habitat and location. From Table 2 in Tarvin and Woolfenden (1999): In IA, 54% in riparian (n=12; Best & Stauffer 1980); In AR, 52% in suburban areas (n=42; Tarvin and Smith 1995); In FL, 39% in suburban (n=31; Rohwer 1969a); In IL, 66% statewide (n=29; Graber et al. 1987); and in FL, 63% in mesic lowland forest (n=13), 17% in shrubby pine flatwoods (n=32), 8% in citrus grove (n=7), 36% in disturbed, shrubby rural habitats (n=30), 21% in disturbed, open rural habitats (n=68), and 11% in xeric upland forest (n=21). Study from MN in Table 2 was omitted because of atypical method of estimating nest success. Weighted mean of above studies is 35% overall nest success. Based on a nest period of 41 d (i.e., 20+18+4-1), estimated daily nest mortality rate is 0.025.
daily mortality rate during nestling-rearing	m2	0.025	
date of first egg of first nest (dd-mmm)	T1	2-Apr	"Examples of first egg dates: Michigan, 5 May (Barrows 1912); Minnesota, 24 Apr (Hilton and Vessall 1980); Kentucky, 14 Apr (Mengel 1965); Kansas, 10 Apr (Johnsgard 1979); Arkansas, 30 Mar (KAT); Alabama, 17 Mar (Imhof 1976); Florida, approximately 7 Mar (Woolfenden and Rohwer 1969a), one full set by 8 Mar (Stevenson and Anderson 1994). Peak laying ranges from early Apr in the South to mid-May in the north" as quoted from Tarvin and Woolfenden (1999). Based on Figure 6 of Tarvin and Woolfenden (1999), assume typical date for start of core egg laying is April 2.
date of first egg of last nest (dd-mmm)	Tlast	16-May	"In native habitats, >1 brood apparently rare. Nicholson (1936) suggested 3 broods/season common in Florida, but Tarvin (1998) found no evidence for >1 successful brood in native habitat in s.-central Florida. Replacement clutches common, however, and may be initiated through Jul." as quoted from Tarvin and Woolfenden (1999). Based on Figure 6 of Tarvin and Woolfenden (1999), assume typical date for end of core egg laying is May 16.
length of rapid follicle growth period (RFG) for each egg (days)	rfg	4	No species-specific data on RFG or egg weight found. RFG for passerines typically 3-4 d.
mean clutch size	clutch	4	"Clutch size 2–7; typically 4–6 in northern populations, 3–4 in Florida (Nicholson 1936, Goodwin 1986, Graber et al. 1987, KAT). Clutch size distribution of 253 nests from (mostly) s. Ontario: 6 (5%), 5 (36%), 4 (31%), 3 (16%), 2 (7%), 1 (6%); mean = 4.00 (Peck 1987). In Illinois, size distribution of 141 clutches: 6 (4%), 5 (63%), 4 (28%), 3 (3%), 2 (1%); mean = 4.66 (Graber et al. 1987). In Arkansas, size distribution of 23 clutches: 5 (35%), 4 (52%), 3 (9%), 2 (4%); mean = 4.17 (KAT). Mean clutch sizes from different samples of nests in Florida: 3.72 (statewide; range 3–5, *n* = 18; Stevenson and Anderson 1994), 3.70 (Tampa/St. Petersburg suburbs; range 3–4, *n* = 10; Woolfenden and Rohwer 1969a), and 3.42 (Archbold Biological Station; range 2–4, *n* = 12; KAT)" as quoted from Tarvin and Woolfenden (1999).
mean intra-egg laying interval (days)	eli	1	1 d (Tarvin 1998).

Description	Code	Value	Notes
egg on which female typically begins incubation–penultimate (1) or last (0)	penult	1	"Daytime incubation presumably begins with laying of penultimate egg, as all eggs in a clutch usually hatch on same day, but female seems to sit on nest beginning with laying of first egg" as quoted from Tarvin and Woolfenden (1999).
duration from start of incubation to hatch (days)	I	18	17–18 d (W. M. Tyler in Bent 1946, Hardy 1961, Gutkin 1978, Hilton and Vesall 1980, Laine 1983, Tarvin and Smith 1995).
duration from hatch to fledging of nestlings (days)	N	20	"Capable of fledging by day 17, and can rapidly run along ground with fluttering hops although not yet volant. Usually fledge at 17–21 d; observed early fledge dates may have been induced by human observers" as quoted from Tarvin and Woolfenden (1999). Assume typical nesting period of 20 d.
duration since nest failure due to other reasons until female initiates new nest (days)	We	7	No specific data found on duration of We. Assume renesting occurs rapidly after nest failure with a typical duration of We of 7 d.
duration since successful fledging until female initiates new nest (days)	Wf	30	"Parents feed offspring for at least 1, sometimes up to 2, mo" as quoted from Tarvin and Woolfenden (1999). No specific data on the duration of Wf. Assume renesting would not occur until first brood reaches independence at approximately one month. Based on the length of the egg laying period defined above, there would not be sufficient time to renest after success.
female body weight (g) during breeding season	BdyWt	87	"Mass of AHY jays from Connecticut averaged 92.4 g ± 7.4 SD (n = 40; Jewell 1986). Mass of AHY jays from Florida averaged 73.7 g (range 59.9–87.1, n = 85; Fisk 1979)" as quoted from Tarvin and Woolfenden (1999). In PA, mean of 86.8 g ± 8.08 (SD; n=462) (Dunning 1984). Assume 87 g is a typical body weight.
diet composition during breeding season		O	Acorns are a staple, but also feed on seeds, some fruits, large invertebrates, and occasionally small vertebrates. Based on Martin et al. (1951), in spring and summer assume 60% seeds and 40% invertebrates for adults. No specific information found on juvenile diet. Since adults regurgitate bolus to juveniles, assume juvenile diet also 60% seeds and 40% invertebrates.
mean number of nest attempts/ female/ season			
mean number of successful broods/ female/ season			"Given low nest success rate (typically <40%) in native habitat, and lengthy nesting cycle (40 d, not including nest construction), few pairs are likely to fledge >1 brood" as quoted from Tarvin and Woolfenden (1999).
mean number of fledglings/ successful nest	fpsn	3.0	Data from Tarvin (1998) on the total number of nests, percent nest success, and the number of fledglings per total nests from six habitat types were used to calculate a weighted mean number of fledglings per successful nest of 3.0.
mean number of fledglings/ female/ season (ARS)	ARS		

Reference List

Barrows, W. B. 1912. Michigan bird life. Michigan Agricultural College, *Spec. Bull. Dept. Zool. Phys.*

Bent, A. C. 1946. Life Histories of North American Crows, Jays, and Titmice. *U.S. Natl. Mus. Bull.* 191.

Best, L. B. and D. F. Stauffer. 1980. Factors affecting nesting success in riparian bird communities. *Condor* 82:149-158.

Dunning, J. B., Jr. 1984. Body weights of 686 species of North American birds. Western Bird Banding Association Monograph No. 1. 38 pp.

Fisk, E. J. 1979. Fall and winter birds near Homestead, Florida. *Bird-Banding* 50:224-303.

Goodwin, D. 1986. Crows of the World. British Mus. (Nat. Hist.) Suffolk, UK.

Graber, J. W., R. R. Graber, and E. L. Kirk. 1987. Illinois birds: Corvidae. Ill. *Nat. Hist. Surv. Biol. Notes* 126:3-17.

Gutkin, R. E. 1978. The reproductive behavior of Blue Jays and an analysis of communal breeding. Master's Thesis. Clemson Univ., Clemson, SC.

Hardy, J. W. 1961. Studies in behavior and phylogeny of certain New World jays (Garrulinae). *Univ. Kans. Sci. Bull.* 42:13149.

Hilton, Jr., B. and J. M. Vesall. 1980. A preliminary report on the breeding behavior of the Blue Jay in Anoka County, Minnesota. *Loon* 52:146-149.

Imhof, T. A. 1976. Alabama birds. Univ. of Alabama Press, University.

Jewell, S. D. 1986. Weights and wing lengths in Connecticut Blue Jays. *Conn. Warbler* 6:47-49.

Johnsgard, P. A. 1979. Birds of the Great Plains. Univ. of Nebraska Press, Lincoln.

Laine, H. 1983. Behavioral aspects of the breeding biology of the Blue Jay (*Cyanocitta cristata*). Phd Thesis. City Univ. of New York, New York.

Martin, A. C., H. S. Zim, and A. L. Nelson. 1951. American wildlife and plants. McGraw-Hill, New York.

Mengel, R. M. 1965. The birds of Kentucky. *Ornithol. Monogr.* 3:1-581.

Nicholson, D. J. 1936. Observations on the Florida Blue Jay. *Wilson Bull.* 48:26-33.

Peck, G. K. 1987. Breeding birds of Ontario: nidiology and distribution. Royal Ontario Mus. Toronto.

Stevenson, H. M. and B. H. Anderson. 1994. The birdlife of Florida. Univ. Press of Florida, Gainesville.

Tarvin, K. A. 1998. The influence of habitat variation on demography of Blue Jays (*Cyanocitta cristata*) in south-central Florida. Phd Thesis. Univ. of South Florida, Tampa.

Tarvin, K. A. and K. G. Smith. 1995. Microhabitat factors influencing predation and success of suburban Blue Jay *Cyanocitta cristata* nests. *J. Avian Biol.* 26:296-304.

Tarvin, Keith A. and Glen E. Woolfenden. 1999. Blue Jay (Cyanocitta cristata), The Birds of North America Online (A. Poole, Ed.). Ithaca: Cornell Lab of Ornithology; Retrieved from the Birds of North America Online: http://bna.birds.cornell.edu.bnaproxy.birds.cornell.edu/bna/species/469.

Woolfenden, G. E. and S. A. Rohwer. 1969a. Breeding birds in a Florida suburb. Bull. Fla. State Mus. 13:1-83.

American crow (*Corvus brachyrhynchos*)	Four-letter Alpha Code: AMCR		
Species life-history parameters	Model Code	Typical value	Rationale
daily mortality rate during laying & incubation	m1	0.012	In CA 63 of 147 nest attempts (43%) fledged young (Caffrey 2000). In British Columbia 13 of 16 nest attempts (81%) fledged young (Campbell et al 1997). In Saskatchewan 45 of 78 nest attempts (58%) fledged young. Weighted mean for overall nest success in these three studies is 50.2%. Over 58 d nest period (36+18+5-1), daily nest survival rate is 0.988, so daily nest mortality rate is 0.012.
daily mortality rate during nestling-rearing	m2	0.012	
date of first egg of first nest (dd-mmm)	T1	28-Mar	Verbeek and Caffrey (2002) summarize peak laying dates as well as earliest and latest laying dates for 6 studies. The mean date for start and end of the peak egg laying period is March 28 through April 8, though crows will renest if nests are destroyed early and the mean date for last egg laid is April 29.
date of first egg of last nest (dd-mmm)	Tlast	19-May	
length of rapid follicle growth period (RFG) for each egg (days)	rfg	5	Assume similar to jackdaw (*Corvus monedula*) (King 1973).
mean clutch size	clutch	5	"Ranges from 3 to 7 eggs, 5 most common" from Verbeek and Caffrey (2002).
mean intra-egg laying interval (days)	eli	1	"Typical rate is 1/d (Black 1941, Emlen 1942, Ignatiuk & Clark 1991), but in some clutches may skip 1-2 d" as quoted in Verbeek and Caffrey (2002).
egg on which female typically begins incubation–penultimate (1) or last (0)	penult	1	"Starts incubation before completion of the clutch (Reimann 1942; usually with laying of third egg (Black 1941) as quoted in Verbeek and Caffrey (2002). Good (1952) states incubation starts with laying of last egg. Assume start with penultimate egg.
duration from start of incubation to hatch (days)	I	18	Verbeek and Caffrey (2002) cites several mean estimates ranging from 16.8 to 23.2 d, but discount the two estimates over 20 d. Most estimates are around 18 d.
duration from hatch to fledging of nestlings (days)	N	36	From Verbeek and Caffrey (2002): In Saskatchewan fledge at 31.5 d ± 1.6 SD (n=71 fledglings; Ignatiuk and Clark 1991). In CA fledge at 38.0 d ± 0.9 SE (n=17 SE). In OK fledge at about 36 d (range 33-43, n=18 fledglings in 8 nests). In MA fledge at 29.6 d (range 16-36, n=15; Chamberlain-Auger et al 1990). When stressed young fledge early. Assume unstressed young, fledge at about 36 d
duration since nest failure due to other reasons until female initiates new nest (days)	We	12	Renests may occur if first nests are disrupted early in the season (Good 1952). In CA first eggs in renest attempt 10 to 12 d after failure (Kilham 1986), but according to Figure 1, other renests may have taken fewer days than that. In BC northwestern crows lay first egg in renests 13.6 ± 1.2 d (n=10) after failure (Butler et al. 1984). Assume typical duration of We is 12 d.
duration since successful fledging until female initiates new nest (days)	Wf	100	American crows have at most one successful nest per season because females care for fledglings for extended period of time. Use 100 d as placeholder for Wf.

	BdyWt		
female body weight (g) during breeding season	477		From Verbeek and Caffrey (2002), IL adult females averaged 474 g (n=245; Black 1941) and OH adult females 491 g (n=45; Hicks & Dambach 1935).
diet composition during breeding season	O		Quantitative analysis by Kalmbach (1939) of 1340 adult stomachs found 72% plant material (54% cultivated and wild seeds and 18% fruit) and 28% animal. Although the animal portion often contains some vertebrates (and their eggs), here we assume it is all invertebrates. Kalmbach (1939) also reported juvenile diets (n=778) consisted of 83.5% animal (primarily invertebrates) and 16.5% vegetable (primarily seeds).
mean number of nest attempts/ female/ season			
mean number of successful broods/ female/ season			
mean number of fledglings/ successful nest	fpsn	2.8	In NY number of fledglings/successful nest 3.06 ± 1.30 SD (n=151) in urban areas and 3.58 ± 1.46 SD (n=33) in rural areas (McGowan 2001). In CA, 1.93 ± 0.11 fledglings/successful nest (n=59; Caffery 2000). In Saskatchewan 2.56 fledglings/successful nest (n=45 nests, Ignatiuk and Clark 1991). Weighted mean = 2.8 fledglings/successful nest
mean number of fledglings/ female/ season (ARS)	ARS		
Note: American crows have at most one successful nest per season because females care for fledglings for extended period of time.			

Reference list

Black, C. T. 1941. Ecological and economic relations of the crow, with special reference to Illinois. Phd Thesis. Univ. of Illinois, Urbana.

Butler, R. W., N. A. M. Verbeek, and H. Richardson. 1984. The breeding biology of the Northwestern Crow. *Wilson Bull.* 96:408-418.

Caffrey, C. 2000. Correlates of reproductive success in cooperatively breeding Western American Crows: if helpers help, it's not by much. *Condor* 102:333-341.

Campbell, R. W., N. K. Dawe, I. McTaggart-Cowan, J. M. Cooper, and G. W. Kaiser. 1997. The birds of British Columbia, Vol. 3: passerines-flycatchers through vireos. R. Br. Columbia Mus. Victoria.

Emlen, Jr., J. T. 1942. Notes on a nesting colony of Western Crows. *Bird-Banding* 13:143-154.

Good, E. E. 1952. The life history of the American Crow *Corvus brachyrhynchos* Brehm. Phd Thesis. Ohio State Univ., Columbus.

Hicks, L. E. and C. A. Dambach. 1935. Sex ratios and weights in wintering Crows. *Bird-Banding* 6:65-66.

Ignatiuk, J. B. and R. G. Clark. 1991. Breeding biology of American crows in Saskatchewan parkland habitat. *Can. J. Zool.* 69:168-175.

Kalmbach, E. R. 1939. The crow in its relation to agriculture. U.S. Dep. Agric. Farmers Bull. no. 1102.

Kilham, L. 1986. Renesting of American crows in Florida and predation by raccoons. *Florida Field Natur.* 14: 21-23.

McGowan, K. J. 2001. Demographic and behavioral comparisons of suburban and rural American Crows (*Corvus brachyrhynchos*). *in* Avian ecology and conservation in an urbanizing world. (Marzluff, J. M., R. Bowman, and R. Donelly, Eds.) Kluwer Acad. Press, Boston, MA.

Reimann, E. J. 1942. Some studies on the Eastern Crow. *Cassinia* 31:19-24.

Verbeek, N. A. and C. Caffrey. 2002. American Crow (*Corvus brachyrhynchos*), The Birds of North America Online (A. Poole, Ed.). Ithaca: Cornell Lab of Ornithology; Retrieved from the Birds of North America Online: http://bna.birds.cornell.edu.bnaproxy.birds.cornell.edu/bna/species/647.

Horned lark (*Eremophila alpestris*) — Four-letter Alpha Code: HOLA

Species life-history parameters	Model Code	Typical value	Rationale
daily mortality rate during laying & incubation	m1	0.021	25% apparent nest success (1 of 4) in Iowa row crops (Patterson and Best 1996); 60% (18 of 30) nest success reported by Pickwell (1931) in Nice (1957); assuming 60% nest success over 24 d nest period = 0.979 daily success rate or 0.021 daily mortality rate 0.0213 daily mortality rate in Kansas (Ricklefs and Bloom 1977).
daily mortality rate during nestling-rearing	m2	0.021	
date of first egg of first nest (dd-mmm)	T1	25-Mar	"Eggs laid from late March to early July in Midwest (Beason 1970), later at higher latitudes." (Beason 1995); assume March 25 to July 5 (i.e., Julian dates 187-85=102);
date of first egg of last nest (dd-mmm)	Tlast	5-Jul	Breeding season length of 3.46 months or 103.8 days in Ricklefs and Bloom 1977
length of rapid follicle growth period (RFG) for each egg (days)	rfg	4	Based on allometric equation (RFG = 2.852*Egg mass ^0.31) from Alisauskas and Ankney (1992), using and egg mass of 2.78 g, assume rfg = 4 d.
mean clutch size	clutch	3	Mean clutch size: 2.5 in British Columbia and WA, 3.0 in CO, 3.2 in IL, 3.3 in Newfoundland, and 3.5 in NWT (Beason 1995); 3.6 in Kansas (Ricklefs and Bloom 1977).
mean intra-egg laying interval (days)	eli	1	One egg laid each day, early in morning (Beason 1995).
egg on which female typically begins incubation–penultimate (1) or last (0)	penult	0	Incubation usually starts when clutch is complete, but in early nests may start with the penultimate egg (Beason 1995).
duration from start of incubation to hatch (days)	I	11	generally 11 d, but may extend to 12 (rarely 13) d for early nests when incubation starts with penultimate egg (Pickwell 1931 and Beason and Franks 1974 as cited in Beason 1995).
duration from hatch to fledging of nestlings (days)	N	10	10 in Illinois and 8-10 in Newfoundland and British Columbia (Beason 1995).
duration since nest failure due to other reasons until female initiates new nest (days)	We	6	Time to renest w/ first egg average 5.8 d (n=5, range 4-8) in Newfoundland (Cannings and Threlfall 1981); "Renesting after nest destruction or desertion was initiated generally within 2 d" in Illinois, though it is unclear whether "initiation" meant new egg laid or nest construction started (Beason and Franks 1974). Unless females maintain fully developed follicles into incubation, it would not be possible to lay an new egg 2 days post failure; Estimate of 6 d in KS from Ricklefs and Bloom (1977) – basis for this estimate is unknown.
duration since successful fledging until female initiates new nest (days)	Wf	7	Renesting after successful fledging occurs about 1 week after young fledge in Illinois and averaged only 4 d after fledging in Newfoundland (Beason 1995); Estimate of 6 d in KS from Ricklefs and Bloom (1977) – basis for this estimate is unknown.
female body weight (g) during breeding season	BdyWt	35	Female wts for 5 subspecies – 37.3 (*E. a. arcticola*), 31.7 (*enthymia*), 39.6 (*hoyti*), 30.5 (*merrilli*), and 36.7 (*praticola*). Mean = 35.2 (from Beason 1995*).

diet composition during breeding season	O	"Adults are strongly granivorous during the breeding season, taking up to 73% seeds, but young are fed exclusively insects" (Beason 1995). Assume adults consume 73% seeds and 27% invertebrates and juveniles consume 100% invertebrates.
mean number of nest attempts/ female/ season		
mean number of successful broods/ female/ season		"In most locations, at least 2 and possibly more successful clutches/yr" (Beason 1995*).
mean number of fledglings/ successful nest	fpsn	2.46	In BC: 30 fledglings from 13 successful nests (mean = 2.31)(Cannings 1981). In IL: 39 fledglings from 15 successful nests (mean = 2.6)(Pickwell 1931). Weighted average = 2.46 fledglings /successful nest.
mean number of fledglings/ female/ season (ARS)	ARS	6.8	6.8 fledglings/female/year (Ricklefs and Bloom 1977).

Reference List

Alisauskas, R. T., and C. D. Ankney. 1992. The cost of egg laying and its relationship to nutrient reserves in waterfowl. In: *Ecology and Management of Breeding Waterfowl.* B. Batt (Ed.), University of Minnesota Press. Pp 30-61.

Beason, R. C. 1970. The annual cycle of the Prairie Horned Lark in west-central Illinois. Master's thesis, Western Illinois Univ., Macomb.

Beason, R. C. 1995. Horned Lark (*Eremophila aplestris*). In the Birds of North America, 195 (A. Poole and F. Gill, Eds.). The Academy of Natural Sciences, Philadelphia, and The American Ornithologists' Union, Washington, DC.

Beason, R. C., and E. C. Franks. 1974. Breeding behavior of the horned lark. *The Auk* 91: 65-74.

Cannings, R. J. 1981. Notes on the nesting of horned larks *Eremophila alpestris* on the Chilcotin Plateau of British Columbia, Canada. *Murrelet* 62: 21-23.

Cannings, R. J., and W. Threlfall. 1981. Horned lark breeding biology at Cape St. Mary's, Newfoundland. *Wilson Bull.* 93, no. 4: 519-30.

Lovell, H. B. 1944. Breeding records of the prairie horned lark in Kentucky. *The Auk* 61: 648-50.

Nice, M. M. 1931. The birds of Oklahoma. *Publ. Univ. Okla. Biol. Surv.* 3.

Patterson, M. P., and L. B. Best. 1996. Bird abundance and nesting success in Iowa CRP fields: the importance of vegetation structure and composition. *Amer. Midl. Natural.* 135: 153-67.

Pickwell, G. B. 1931. The Prairie Horned Lark. *St. Louis Acad. Sci. Trans.* 27: 1-153.

Ricklefs, R. E., and G. Bloom. 1977. Components of avian breeding productivity. *The Auk* 94: 86-96.

Verbeek, N. A. M. 1967. Breeding biology and ecology of the horned lark in alpine tundra. *Wilson Bull.* 79, no. 2: 208-18.

Tree swallow (*Tachycineta bicolor*) (Northern range)

Four-letter Alpha Code: TRES

Species life-history parameters	Model Code	Typical value	Rationale
daily mortality rate during laying & incubation	m1	0.0056	Summary of 3458 nests from 10 populations had overall nest success (i.e., nests producing ≥1 fledgling) of 78.8% ± 17.2 (SD) (Robertson et al. 1992). Four additional studies with 6301 nest had overall nest success of 82% (range 65 to 100%) (Winkler et al. 2011). Assume 80% nest success over 40 day nesting period (i.e., 20+15+6-1) which is equivalent to a 0.9944 daily nest success rate, or 0.0056 daily nest mortality rate.
daily mortality rate during nestling-rearing	m2	0.0056	
date of first egg of first nest (dd-mmm)	T1	5-May	"Typically lay eggs throughout May and June, but substantial variation depending on location. Further north, breeding commences later. In New York, median lay date of 10 May (Ardia 2007); in Alaska, 1 Jun (Ardia 2007). In Nova Scotia, all breeding attempts started between 12 May and 8 Jun (Thomas and Shutler 2001). In Ontario (Fig. 2), most eggs laid 5 to 22 May; range 3 May to 1 Jul" as quoted from Winkler et al. (2011). Assume laying starts May 5 and renests are possible thru end of June.
date of first egg of last nest (dd-mmm)	Tlast	30-Jun	
length of rapid follicle growth period (RFG) for each egg (days)	rfg	4	Based on allometric equation (RFG = 2.852*Egg mass ^0.31) from Alisauskas and Ankney (1992), using and egg mass of 2.6 g, assume rfg = 4 d.
mean clutch size	clutch	6	Mean of 28 studies summarized in Winkler et al. (2011) was 5.59 eggs/clutch.
mean intra-egg laying interval (days)	eli	1	1 egg/d (Robertson et al. 1992).
egg on which female typically begins incubation–penultimate (1) or last (0)	penult	1	"Females usually begin incubation on day penultimate egg is laid" as quoted in Robertson et al. (1992).
duration from start of incubation to hatch (days)	I	15	"11 to 19 d range, but usually 14 to 15 d (Austin and Low 1932, Kuerzi 1941, Paynter 1954, Stocek 1970)" as quoted in Robertson et al. (1992).
duration from hatch to fledging of nestlings (days)	N	20	"The period from hatching to departure from nest ranges from 15 to 25 d, although most nestlings depart nest at an age of 18 to 22 d (Austin and Low 1932, Kuerzi 1941, Paynter 1954, Burtt 1977)" as quoted in Robertson et al. (1992). Assume 20d is typical nestling period.
duration since nest failure due to other reasons until female initiates new nest (days)	We	10	"Females often lay a second clutch within 10 d In response to abandonment or loss of first clutch (Kuerzi 1941, Paynter 1954, Tyler 1942)" as quoted in Robertson et al. (1992).
duration since successful fledging until female initiates new nest (days)	Wf	100	"Second broods (as opposed to renests) rare (Chapman 1955, Hussell 1983a), except perhaps in southern part of range" as quoted in Robertson et al. (1992). Assume females quit breeding after success, so set Wf to default of 100.
female body weight (g) during breeding season	BdyWt	20.7	Females in April 20.7 g ± 1.3 (m=36, range 18.0-23.0 g) (Robertson et al. 1992).

Species Life-History Profiles – 12 December 2013

diet composition during breeding season	I		"During the breeding season, Tree Swallows rely almost entirely on aerial insects captured during prolonged cruising flights both to feed themselves and their dependent offspring" as quoted from McCarty and Winkler (1999). Assume adult and juvenile diets consist of 100% insects.
mean number of nest attempts/ female/ season			
mean number of successful broods/ female/ season			"Second broods (as opposed to renests) rare (Chapman 1955, Hussell 1983a), except perhaps in southern part of range" as quoted in Robertson et al. (1992).
mean number of fledglings/ successful nest	fpsn	4.5	Summary of 8 populations produced 13,137 fledglings from 2821 successful nest for mean of 4.66 (Robertson et al. 1992). Four additional studies produced approx. 19,450 fledglings from 4529 successful nests for mean of 4.29 (Winkler et al. 2011). Assume a mean of 4.5 fledglings/successful nest.
mean number of fledglings/ female/ season (ARS)	ARS		

Reference list

Alisauskas, R. T., and C. D. Ankney. 1992. The cost of egg laying and its relationship to nutrient reserves in waterfowl. In: Ecology and Management of Breeding Waterfowl. B. Batt (Ed.), University of Minnesota Press. Pp 30-61.

Ardia, D. R. 2007. Site- and sex-level differences in adult feeding behaviour and its consequences to offspring quality in Tree Swallows (*Tachycineta bicolor*) following brood-size manipulation. *Can. J. Zool.-Rev. Canad. Zool.* 85:847-854.

Austin, O. L. and S. H. Low. 1932. Notes on the breeding of the Tree Swallow. *Bird-Banding* 3:39-44.

Burtt, Jr., E. H. 1977. Some factors in the timing of parent chick recognition in swallows. *Anim. Behav.* 25:231-239.

Chapman, L. B. 1955. Studies of a Tree Swallow colony. *Bird-Banding* 26:45-70.

Hussell, D. J. T. 1983. Tree Swallow pairs raise two broods per season. *Wilson Bull.* 95:470-471.

Kuerzi, R. G. 1941. Life history studies of the Tree Swallow. *Proc. Linn. Soc. N.Y.* 52-53:1-52.

McCarty, J. P. and D. W. Winkler. 1999. Foraging ecology and diet selectivity of Tree Swallows feeding nestlings. *Condor* 101(2):246-254.

Paynter, R. A. 1954. Interrelationships between clutch-size, brood-size, prefledging survival and weight in Kent Island Tree Swallows. *Bird-Banding* 25:35-58, 102-110, 136-148.

Robertson, R. J., B. J. Stutchbury, and R. R. Cohen. 1992. Tree Swallow. In The Birds of North America, No. 11 (A. Poole, P. Stettenheim, and F. Gill, Eds.). Philadelphia: The Academy of Natural Sciences: Washington, DC: The American Ornithologists' Union.

Stocek, R. F. 1970. Observations on the breeding ecology of the Tree Swallow. *Cassinia* 52:3-20.

Winkler, D. W., K. K. Hallinger, D. R. Ardia, R. J. Robertson, B. J. Stutchbury and R. R. Cohen. 2011. Tree Swallow (*Tachycineta bicolor*), The Birds of North America Online (A. Poole, Ed.). Ithaca: Cornell Lab of Ornithology; Retrieved from the Birds of North America Online: http://bna.birds.cornell.edu.bnaproxy.birds.cornell.edu/bna/species/011.

Tree swallow (*Tachycineta bicolor*) (Southern range)			Four-letter Alpha Code: TRES
Species life-history parameters	Model Code	Typical value	Rationale
daily mortality rate during laying & incubation	m1	0.0056	Summary of 3458 nests from 10 populations had overall nest success (i.e., nests producing ≥ 1 fledgling) of 78.8% $\pm$ 17.2 (SD) (Robertson et al. 1992). Four additional studies with 6301 nest had overall nest success of 82% (range 65 to 100%) (Winkler et al 2011). Assume 80% nest success over 40 day nesting period (i.e., 20+15+6-1) which is equivalent to a 0.9944 daily nest success rate, or 0.0056 daily nest mortality rate.
daily mortality rate during nestling-rearing	m2	0.0056	
date of first egg of first nest (dd-mmm)	T1	19-Apr	"Typically lay eggs throughout May and June, but substantial variation depending on location. In southern extremes of breeding range, clutch initiation may begin as early as April. First egg date 19 Apr in Virginia (D. A. Cristol, pers. comm.); median lay date 21 Apr in Tennessee (Ardia and Rice 2006). ....egg laying for second broods is typically concurrent with that of late-breeding females laying for the first time or laying replacement clutches after the failure of their first (e.g. Virginia: 7-26 Jun; Monroe et al. 2008)" as quoted in Winkler et al. (2011). Assume laying first eggs between April 19 and 26 June.
date of first egg of last nest (dd-mmm)	Tlast	26-Jun	
length of rapid follicle growth period (RFG) for each egg (days)	rfg	4	Based on allometric equation (RFG = 2.852*Egg mass ^0.31) from Alisauskas and Ankney (1992), using and egg mass of 2.6 g, assume rfg = 4 d.
mean clutch size	clutch	6	Mean of 28 studies summarized in Winkler et al. (2011) was 5.59 eggs/clutch.
mean intra-egg laying interval (days)	eli	1	1 egg/d (Robertson et al. 1992).
egg on which female typically begins incubation—penultimate (1) or last (0)	penult	1	"Females usually begin incubation on day penultimate egg is laid" as quoted in Robertson et al. (1992).
duration from start of incubation to hatch (days)	I	15	"11 to 19 d range, but usually 14 to 15 d (Austin and Low 1932, Kuerzi 1941, Paynter 1954, Stocek 1970)" as quoted in Robertson et al. (1992).
duration from hatch to fledging of nestlings (days)	N	20	"The period from hatching to departure from nest ranges from 15 to 25 d, although most nestlings depart nest at an age of 18 to 22 d (Austin and Low 1932, Kuerzi 1941, Paynter 1954, Burtt 1977)" as quoted in Robertson et al. (1992). Assume 20d is typical nestling period.
duration since nest failure due to other reasons until female initiates new nest (days)	We	10	"Females often lay a second clutch within 10 d In response to abandonment or loss of first clutch (Kuerzi 1941, Paynter 1954, Tyler 1942)" as quoted in Robertson et al. (1992).
duration since successful fledging until female initiates new nest (days)	Wf	9	"Double-brooding females waited, on average, 9.3 $\pm$ 3.2 days after young from their first nest had fledged before initiating a second clutch (range: 3–15 days)" as quoted in Monroe et al. (2008).
female body weight (g) during breeding season	BdyWt	20.7	Females in April 20.7 g $\pm$ 1.3 (n=36, range 18.0-23.0 g) (Robertson et al. 1992).

diet composition during breeding season	I		"During the breeding season, Tree Swallows rely almost entirely on aerial insects captured during prolonged cruising flights both to feed themselves and their dependent offspring" as quoted from McCarty and Winkler (1999). Assume adult and juvenile diets consist of 100% insects.
mean number of nest attempts/ female/ season			
mean number of successful broods/ female/ season			
mean number of fledglings/ successful nest	fpsn	4.5	Summary of 8 populations produced 13,137 fledglings from 2821 successful nest for mean of 4.66 (Robertson et al. 1992). Four additional studies produced approx. 19,450 fledglings from 4529 successful nests for mean of 4.29 (Winkler et al. 2011). Assume a mean of 4.5 fledglings/successful nest.
mean number of fledglings/ female/ season (ARS)	ARS		

Reference list

Alisauskas, R. T., and C. D. Ankney. 1992. The cost of egg laying and its relationship to nutrient reserves in waterfowl. In: Ecology and Management of Breeding Waterfowl. B. Batt (Ed.), University of Minnesota Press. Pp 30-61.

Ardia, D. R. and E. B. Rice. 2006. Variation in heritability of immune function in the Tree Swallow. Evol. Ecol. 20(5):491-500.

Austin, O. L. and S. H. Low. 1932. Notes on the breeding of the Tree Swallow. Bird-Banding 3:39-44.

Burtt, Jr., E. H. 1977. Some factors in the timing of parent chick recognition in swallows. Anim. Behav. 25:231-239.

Chapman, L. B. 1955. Studies of a Tree Swallow colony. Bird-Banding 26:45-70.

Hussell, D. J. T. 1983. Tree Swallow pairs raise two broods per season. Wilson Bull. 95:470-471.

Kuerzi, R. G. 1941. Life history studies of the Tree Swallow. Proc. Linn. Soc. N.Y. 52-53:1-52.

McCarty, J. P. and D. W. Winkler. 1999. Foraging ecology and diet selectivity of Tree Swallows feeding nestlings. Condor 101(2):246-254.

Monroe, A. P., K. K. Hallinger, R. L. Brasso, and D. A. Cristol. 2008. Occurrence and implications of double brooding in a southern population of Tree Swallows. Condor 110(2):382-386.

Paynter, R. A. 1954. Interrelationships between clutch-size, brood-size, prefledging survival and weight in Kent Island Tree Swallows. *Bird-Banding* 25:35-58, 102-110, 136-148.

Robertson, R. J., B. J. Stutchbury, and R. R. Cohen. 1992. Tree Swallow. In The Birds of North America, No. 11 (A. Poole, P. Stettenheim, and F. Gill, Eds.). Philadelphia: The Academy of Natural Sciences: Washington, DC: The American Ornithologists' Union.

Stocek, R. F. 1970. Observations on the breeding ecology of the Tree Swallow. *Cassinia* 52:3-20.

Winkler, David W., Kelly K. Hallinger, Daniel R. Ardia, R. J. Robertson, B. J. Stutchbury and R. R. Cohen. 2011. Tree Swallow (*Tachycineta bicolor*), The Birds of North America Online (A. Poole, Ed.). Ithaca: Cornell Lab of Ornithology; Retrieved from the Birds of North America Online: http://bna.birds.cornell.edu.bnaproxy.birds.cornell.edu/bna/species/011.

Barn swallow (*Hirundo rustica*) — **Four-letter Alpha Code: BARS**

(BARS) life-history parameters	Model Code	Typical value	Rationale
daily mortality rate during laying & incubation	m1	0.0186	Shields and Crook (1987) reported the percentage of nest attempts that fledged ≥ 1 chick separately for 4 colony sizes and for 1st and 2nd clutches. Weighted mean for all nests (n=190) gives an overall nest success of 48.95%. Assuming nest period of 38 d (4 d for egg laying + 14 d incubation + 20 d nestlings), the estimated daily survival rate is 0.9814. Daily nest mortality rate is 1-0.9814= 0.0186.
daily mortality rate during nestling-rearing	m2	0.0186	
date of first egg of first nest (dd-mmm)	T1	1-May	"In Kansas, clutch initiation dates spanned 1 May-10 Aug in one study, with most first clutches started 15-25 May (n=104; Johnston 1964)." Modal clutch initiation date for second clutches was 5 Jul in KS (Johnston1964). In a second study in Kansas, Anthony and Ely (1976) found first clutches peaked 1-5 Jun while second clutches peaked 15-16 Jul.
date of first egg of last nest (dd-mmm)	Tlast	10-Aug	
length of rapid follicle growth period (RFG) for each egg (days)	rfg	4	Based on allometric equation from Alisauskas and Ankney (1992), assume rfg = 4 d based on 2.0 g egg.
mean clutch size	clutch	5	In KS mean 4.7 eggs (mode 5, range 3-7, n=43), with average clutch size declining by $\sim$ 0.6 egg from May1 to Aug 10 (Johnston 1964); also in KS first clutch mean is 4.6 (range 4-7, n=105) and second clutch is 4.1 eggs (range 3-6, n=68) (Anthony and Ely 1976): In MI mean 4.7 ± 0.9 SD (range 3-6, n=17) (Goodman 1982): In WV 1st clutch 4.6 ± 0.1 SD (n=94) and 2nd 4.1 ± 0.3 SD (n=33) (Samuel 1971): weighted mean = 4.5.
mean intra-egg laying interval (days)	eli	1	1 egg laid per day in early morning (Anthony and Ely 1976 from Brown and Brown 1999).
egg on which female typically begins incubation–penultimate (1) or last (0)	penult	1	"Incubation is thought to begin after laying of penultimate egg in Vermont (Smith 1933) and Kansas (Thompson 1961), and after third egg of 5 in Kentucky (Tabler 1956)." (Brown and Brown 1999).
duration from start of incubation to hatch (days)	I	14	"Reported as 13.8 (range 13-15, n=20) and 13.7 (n=118) in KS (Thompson 1961, Anthony and Ely 1976); 14.6 (range 12-17, n=42) in BC (Campbell et al. 1997)" (Brown and Brown 1999); assume 14 d.
duration from hatch to fledging of nestlings (days)	N	20	"Fledge at mean of 20.4 and 20.7 d (1st & 2nd broods, resp.) of age in KS (range 15-24 d; Anthony and Ely 1976); In WV, 20.7 d (range 18-27, n=10; Samuel 1971); In BC, 19.5 d (range 19-24, n=12; Campbell et al. 1997)" (Brown and Brown 1999); assume 20 d.
duration since nest failure due to other reasons until female initiates new nest (days)	We	11	No info specifically on duration til renesting after failure. Assuming that failures result in building of new nests, the mean time for building new nest was 11.0 in NY (Shields et al. 1988), 6.4 d in WV (Samuel 1971), and 15 d in BC (Campbell et al 1997) for mean of ~11 d.

duration since successful fledging until female initiates new nest (days)	Wf	15	"Juveniles are fed by their parents for several days, possibly up to a week, after fledging; family groups have broken up entirely by 2 wk after fledging (Smith 1937, Medvin and Beecher 1986)" (Brown and Brown 1999). Assume females spend 7 days caring for young post-fledging. Once first clutch of juveniles leave, female may renest by building new nest or remodeling old nest. In Manitoba, 12% (6/50) of birds used same nest for 1st and 2nd clutches (Barclay 1988), whereas in OK the percentage was 81.3% (n=16) (Iverson 1988) and KS 45.2% (n=137) (Anthony and Ely 1976). In NY 53% reused first nest (Shields 1984). "In New York, birds reusing an old nest spent mean of 4.8 d refurbishing it (n=45), compared to 11.0 d building new nests (n=15, Shields et al. 1988)." Assume equal split between reusing old nest vs new nest for an average of about 8 d to prepare nest site. So, 7 + 8 = 15.
female body weight (g) during breeding season	BdyWt	19.2	19.2 ± 1.5 SD (n=130) (Brown and Brown 1999).
diet composition during breeding season		I	Adults consume almost exclusively flying insects throughout the year (Brown and Brown 1999). Juveniles fed bolus of regurgitated insects from both parents (Brown and Brown 1999). Assume both adults and juveniles consume 100% invertebrates.
mean number of nest attempts/ female/ season			"Often 2. In ON, 30% of females initiated second clutch (Smith and Montgomerie 1991); In Manitoba, 90% (Barclay 1988); In MA, 67% (n=52; Mason 1953); In NY, 49% (n=301; Snapp 1976)" (Brown and Brown 1999).
mean number of successful broods/ female/ season			
mean number of fledglings/ successful nest	fpsn	3.67	In MA, 4.3 young/successful nest from first brood (n=374) and 3.3 young/successful nest from second brood (n=101) (Mason 1953). weighted average = 4.087 young or 2.04 female young/successful nest. In MS, 3.4 (n=222) and 3.1 (n=235) fledglings/nest during two years (Lohoefener 1980). Weighted average = 3.25 fledglings or 1.62 female fledglings/nest. Weighted average for MA & MS = 3.67 fledglings/ successful nest.
mean number of fledglings/ female/ season (ARS)	ARS	5.53	Mean ARS in Manitoba = 6.86 ± 0.4 SD (range 3-11, n=21) fledglings/pair (Barclay 1988) and in ON = 4.2 fledglings/pairs (n=20) (Smith and Montgomerie 1991). Weighted mean = 5.53 fledglings/female/year. Proportion of females that are nonbreeders is unknown (Brown and Brown 1999).

Reference List

Alisauskas, R. T., and C. D. Ankney. 1992. The cost of egg laying and its relationship to nutrient reserves in waterfowl. In: *Ecology and Management of Breeding Waterfowl.* B. Batt (Ed.), University of Minnesota Press. Pp 30-61.

Anthony, L. W. and C. A. Ely. 1976. Breeding biology of barn swallows in west-central Kansas. *Bull. Kans. Ornithol. Soc.* 27:37-43.

Barclay, R. M. R. 1988. Variation in the costs, benefits, and frequency of nest reuse by barn swallows (*Hirundo rustica*). *The Auk* 105: 53-60.

Brown, C. R. and M. B. Brown. 1999. Barn Swallow (*Hirundo rustica*). In Birds of North America, No. 452 (A. Poole and F. Gill, Eds.). The Birds of North America, Inc., Philadelphia, PA.

Campbell, R. W., N. K. Dawe, I. McTaggart-Cowan, J. M. Cooper, G. W. Kaiser, et al. 1997. The birds of British Columbia, Vol. 3. University of British Columbia Press, Vancouver.

Goodman, S. M. 1982. A test of nest cup volume and reproductive success in the barn swallow. *Jack-Pine Warbler* 60:107-112.

Johnson, R. F. 1964. The breeding birds of Kansas. *University of Kansas Publ. Mus. Nat. Hist.* 12:575-655.

Lohoefener, R. R. 1980. Comparative breeding biology and ethology of colonial and solitary nesting barn swallows (*Hirundo rustica*) in east-central Mississippi. PhD diss., Mississippi State University, Starkville, MS.

Mason, E. A. 1953. Barn swallow life history data based on banding records. *Bird Banding* 24, no. 3: 91-100.

Medvin, M. B. and M. D. Beecher. 1986. Parent-offspring recognition in the barn swallow (*Hirundo rustica*). *Anim. Behav.* 34:1627-1639.

Samuel, D. E. 1971. The breeding biology of barn and cliff swallows in West Virginia. *Wilson Bull.* 83, no. 3: 284-300.

Shields, W. M. 1984. Factors affecting nest and site fidelity in Adirondack barn swallows (*Hirundo rustica*). *The Auk* 101: 780-789.

Shields, W. M., and J. R. Crook. 1987. Barn swallow coloniality: A net cost for group breeding in the Adirondacks? *Ecology* 68, no. 5: 1373-86.

Shields, W. M., J. R. Crook, M. L. Hebblethwaite, and S. S. Wiles-Ehmann. 1988. Ideal free coloniality in the swallows. Pp. 189-228 in The ecology of social behavior (C. N. Slobodchikoff, Ed.). Academic Press, San Diego, CA.

Smith, W. P. 1937. Further notes on the nesting of the barn swallow. *The Auk* 54:65-69.

Smith, H. G and R. Montgomerie. 1991. Sexual selection and the tail ornaments of North American barn swallows. *Behav. Ecol. Sociobiol.* 28:195-201.

Snapp, B. D. 1976. Colonial breeding in the barn swallow (*Hirundo rustica*) and its adaptive significance. *The Condor* 78: 471-80.

Carolina chickadee (*Poecile carolinensis*) Four-letter Alpha Code: CACH

Species life-history parameters	Model Code	Typical value	Rationale
daily mortality rate during laying & incubation	m1	0.0144	Calculated in Table 6 of Albano (1992).
daily mortality rate during nestling-rearing	m2	0.0104	Calculated in Table 6 of Albano (1992).
date of first egg of first nest (dd-mmm)	T1	2-Apr	Mean of first laying dates in Table 1 of Mostrom et al. 2002 (April 2).
date of first egg of last nest (dd-mmm)	Tlast	14-May	Mean of last laying dates in Table 1 of Mostrom et al. 2002 (May 14).
length of rapid follicle growth period (RFG) for each egg (days)	rfg	3	Based on allometric equation from Alisauskas and Ankney (1992), assume rfg = 3 d based on 1.04 g egg.
mean clutch size	clutch	6	Mean of 5.8 eggs/clutch and mode of 6 for compiled list of sources in Table 1 of Mostrom et al. (2002).
mean intra-egg laying interval (days)	eli	1	Typically 1 egg laid each day. Single day skipped rarely (Mostrom et al. 2002)
egg on which female typically begins incubation–penultimate (1) or last (0)	penult	0	Regular incubation begins with laying of last, or occasionally penultimate, egg (Brewer 1961 as cited in Mostrom et al. 2002).
duration from start of incubation to hatch (days)	I	13	Average of 12.9 d ± 0.4 SE (range 12-15 d; n=12; Albano 1992).
duration from hatch to fledging of nestlings (days)	N	17	Average incubation period of 16.8 d ± 0.2 (n=25) (Albano 1992); 16-19 d after hatch (Pitts 1978, 1997).
duration since nest failure due to other reasons until female initiates new nest (days)	We	10	No specific info found on time to renesting after nest failure for Carolina chickadee; assume 10 d based on information for black-capped chickadees.
duration since successful fledging until female initiates new nest (days)	Wf	28	Both parents feed fledglings. Become independent of parents within 2-3 weeks (Brewer 1961, Pitts 1997); assume 21 d to care for fledglings plus 7 d to create new nest = 28 d.
female body weight (g) during breeding season	BdyWt	10	In FL 9.8 g ± 0.1 SD (n=24, Hartman 1961); in MO 10.8 g ± 0.6 SD (n=8, Robbins et al. 1986); in IL approx. 9.8 g (Brewer 1961); in PA 9.9 g ± 0.5 SD (n=30, Mostrom et al. 2002).
diet composition during breeding season		O	Throughout the year, consume 72% invertebrates and 28% fruits and seeds (Howell 1932 cited in Mostrom et al. 2002), but more invertebrates during the breeding season. During the breeding season, diet similar to the black-capped chickadee (Martin et al. 1951), so assume adults feed 90% invertebrates, 5% fruits, and 5% seeds and juveniles fed 100% invertebrates.

Parameter	Abbrev	Value	Notes
mean number of nest attempts/ female/ season		1	Replacement clutches and second-brood attempts are rare (Mostrom et al. 2002).
mean number of successful broods/ female/ season		0.657	Assuming that replacement clutches and second-brood attempts are rare and that all females initiate a clutch, number of successful nests/female would be equal to overall nest success—Mayfield-corrected nest success for 1989 & 1990 was 65.7% (Albano 1992).
mean number of fledglings/ successful nest	fpsn	3.9	Albano (1992) reports that 40 of 51 nests fledged at least one chick, but Table 5 indicates that the number of chicks fledged averaged 3.9 ± 0.4 from 44 nests, so the average may be based on some unsuccessful nests, but this is unclear. Consequently, 3.9 fledglings/successful nest may slightly underestimate the number of fledglings/successful nest, but it is the best estimate available.
mean number of fledglings/ female/ season (ARS)	ARS	2.6	$0.657 * 3.9 = 2.6$ Assume approx. 2.6 fledglings/female/year.

Reference List

Alisauskas, R. T., and C. D. Ankney. 1992. The cost of egg laying and its relationship to nutrient reserves in waterfowl. In: *Ecology and Management of Breeding Waterfowl*. B. Batt (Ed.), University of Minnesota Press. Pp 30-61.

Albano, D. J. 1992. Nesting mortality of Carolina chickadees breeding in natural cavities. *The Condor* 94: 371-82.

Brewer, R. 1961. Comparative notes on the life history of the Carolina chickadee. *Wilson Bull.* 73: 348-73.

Martin, A. C., H. S. Zim, and A. L. Nelson. 1951. *American Wildlife and Plants: A Guide to Wildlife Food Habits*. Dover Publications, Inc., New York. 500 pp.

Mostrom, A. M., R. L. Curry, and B. Lohr. 2002. Carolina Chickadee (*Poecile carolinensis*). In Birds of North America, No. 636 (A. Poole and F. Gill, Eds.). The Birds of North America, Inc., Philadelphia, PA.

Pitts, T. D. 1978. Some nesting habits of Carolina chickadees. *J. Tenn. Acad. Sci.* 53:14-16.

Pitts, T. D. 1997. Carolina chickadee. Pp. 219-221 in Atlas of the breeding birds of Tennessee (C. P. Nicholson, Ed.). University of Tennessee Press, Knoxville.

Black-capped chickadee (*Poecile atricapillus*)			Four-letter Alpha Code: BCCH
Species life-history parameters	Model Code	Typical value	Rationale
daily mortality rate during laying & incubation	m1	0.013	Many studies report hatching success and fledging success, but not nest success. 7 nests successful out of 11 (64%) reported by Odum (1941) in New York and 17 nests successful out of 25 (68%) by Kluyver (1961) in MA. 64% over a 34 d nest period is a daily survival rate of 0.987 (i.e., 0.987^34 = 0.64); 1-0.987 = 0.013 daily mortality rate; 0.0090 daily mortality rate (Ricklefs and Bloom 1977).
daily mortality rate during nestling-rearing	m2	0.013	
date of first egg of first nest (dd-mmm)	T1	23-Apr	Egg dates: Illinois April 20-June 11 (n=17), Massachusetts May 7-Julu 12 (n=27); Nova Scotia May 21-June 6 (n=5); Oregon April 13-June 30 (n=57; Bent 1946); Michigan April 18-July 3 (n=38; Nickell 1956); Average of U.S. dates (Nova Scotia window seemed very narrow so was excluded from calculations) are April 23 to June 29 (Julian dates 180-113=68); 2.20 months or 66 days (Ricklefs and Bloom 1977).
date of first egg of last nest (dd-mmm)	Tlast	29-Jun	
length of rapid follicle growth period (RFG) for each egg (days)	rfg	3	Assume rfg is same as estimated for Carolina chickadee (i.e., 3 d).
mean clutch size	clutch	7	80% of 225 clutches has 6, 7, or 8 eggs (Smith 1991); 5.4 (Ricklefs and Bloom 1977).
mean intra-egg laying interval (days)	eli	1	Typically 1 egg per d (Smith 1993).
egg on which female typically begins incubation–penultimate (1) or last (0)	penult	1	Incubation starts with penultimate egg (Smith 1993).
duration from start of incubation to hatch (days)	I	12	Incubation period usually 12-13 d (Smith 1993).
duration from hatch to fledging of nestlings (days)	N	16	Usually 16 d, but if disturbed, as early as 12 d (Smith 1993).
duration since nest failure due to other reasons until female initiates new nest (days)	We	10	Odum (1941b) reported first egg in new nest 10 or 11 d after first nest destroyed. Birds need time to excavate new nest. "Replacement broods, begun after loss of a first brood, often started within a few days of loss" (Smith 1993). Ricklefs and Bloom (1977) used 10 d after failure.
duration since successful fledging until female initiates new nest (days)	Wf	35	"Young stay with adults for 3-4 wk (occasionally as little as 2) after leaving the nest. To begin with, parents provide all of their food, although young usually catch first food items with a week of leaving the nest " (Smith 1993); additional time would be required to construct a new cavity, so assume 25 d of nestling care plus 10 d for nest construction and egg formation. Ricklefs and Bloom (1977) used 10 d after success, but this ignores evidence that females participate in caring for fledglings for several weeks.
female body weight (g) during breeding season	BdyWt	10.8	10.8 ± 1.38 (SD), n=1880 from Dunning (1984).

Species Life-History Profiles – 12 December 2013

diet composition during breeding season	O	In breeding season, mostly caterpillars, along with a variety of other insects, spiders, small snails, slugs and centipedes; also berries from many species of plants (Smith 1993). "Summer: about 80% - 90% animal, 10% - 20% plant" (Martin et al. 1951 cited in Smith 1993). Figures in Martin et al. 1951 indicate about 10% plant material during breeding season. Juveniles fed primarily invertebrates, mainly caterpillars (Smith 1993). Assume adults feed on 90% invertebrates, 5% fruits, and 5% seeds and juveniles fed 100% invertebrates.
mean number of nest attempts/ female/ season	1	Usually 1 clutch per year; second broods are rare (Smith 1993). Kluyver (1961) reported 1 of 22 pairs undertook a second pair and 2 of 22 pairs attempted a repeat nest after first was disrupted.
mean number of successful broods/ female/ season		
mean number of fledglings/ successful nest	fpsn	5.3	Number of fledglings/ successful nest: 6.6 (Nickell 1956); 4.2 (Kluyver 1961); 4.0 and 4.6 (Smith 1967) (as cited in Smith 1993), however, based on data in Kluyver (1961) this should be 5.0 fledglings/successful nest (85/17). Average of three studies (6.6+5.0+4.3)/3 = 5.3 fledglings/successful nest.
mean number of fledglings/ female/ season (ARS)	ARS	6.15	6.15 fledglings/female/year (Ricklefs and Bloom 1977).

Reference List

Bent, A. C. 1946. Life histories of North American jays, crows and titmice. *U.S. National Museum Bulletin* 191.

Dunning, J. B. Jr. 1984. *Body weights of 686 species of North American birds,*, Eldon Publishing, Cave Creek, AZ.

Kluyver, H. N. 1961. Food consumption in relation to habitat in breeding chickadees. *The Auk* 78: 532-50.

Martin, A. C., H. S. Zim, and A. L. Nelson. 1951. *American Wildlife and Plants: A Guide to Wildlife Food Habits.* Dover Publications, Inc., New York. 500 pp.

Nickell, W. P. 1956. Nesting of the black-capped chickadee in the southern peninsula of Michigan. *Jack-Pine Warbler* 34:317-138.

Odum, E. P. 1941. Annual cycle of the black-capped chickadee --2. *The Auk* 58: 518-35.

Ricklefs, R. E., and G. Bloom. 1977. Components of avian breeding productivity. *The Auk* 94: 86-96.

Smith, S. M. 1967. Seasonal changes in the survival of the black-capped chickadee. *Condor* 69:344-359.

Smith, S. M. 1991. The black-capped chickadee: behavioral ecology and natural history. Cornell University Press, Ithaca, NY.

Smith, S. M. 1993. Black-capped Chickadee (*Parus atricapillus*). In Birds of North America, No. 39 (A. Poole, P. Stettenheim, and F. Gill, Eds.). Philadelphia, The Academy of Natural Sciences; Washington, DC: The Ornithologists' Union.

Verdin (*Auriparus flaviceps*) — Four-letter Alpha Code: VERD

Species life-history parameters	Model Code	Typical value	Rationale
daily mortality rate during laying & incubation	m1	0.0106	Overall mean nest success for all studies (i.e., TX, NM AZ, NV) reported in Table 4 of Austin (1977) was 67.4%. Daily mortality rates from laying to fledging averaged 1.06% (0.39-2.97) for nests (Austin 1977).
daily mortality rate during nestling-rearing	m2	0.0106	
date of first egg of first nest (dd-mmm)	T1	Apr-7	Based on the core egglaying period shown in Figure 4 in Webster (1999) and in the compilation of new nest initiations per month from multiple sites shown in Table 2 in Austin (1977), assume the typical dates for initiating egg laying are from April 7 to June 25.
date of first egg of last nest (dd-mmm)	Tlast	Jun-25	
length of rapid follicle growth period (RFG) for each egg (days)	rfg	3	Based on allometric equation (RFG = 2.852*Egg mass ^0.31) from Alisauskas and Ankney (1992), using and egg mass of 0.96 g, assume rfg = 3 d.
mean clutch size	clutch	4	"Early clutches typically larger (average 3.86 eggs in Mar–Apr [n = 131 nests] versus 3.28 eggs in Jun–Jul nests; n = 146 nests)" as quoted in Webster (1999).
mean intra-egg laying interval (days)	eli	1	"Lays eggs singly on succeeding days, early in morning. Female does not attend nest during egg-laying period" as quoted in Webster (1999).
egg on which female typically begins incubation–penultimate (1) or last (0)	penult	0	"Female does not attend nest during egg-laying period" as quoted in Webster (1999).
duration from start of incubation to hatch (days)	I	16	"Lasts 14–18 d (Hensley 1959, Taylor 1971)" as quoted in Webster (1999).
duration from hatch to fledging of nestlings (days)	N	18	Most birds fledge by day 18 (average mass 7.0 g, range 6.5–7.6, n = 8) (Webster 1999).
duration since nest failure due to other reasons until female initiates new nest (days)	We	4	After a successful nesting, an average of 2 days elapsed between fledging and the first egg of the next clutch; after a failure, an average of 4 days elapsed (Taylor 1967, Austin 1977). For Wf assume 3 days since in the basic version of MCnest, females are not allowed to have an overlap between the fledgling care phase and the rfg period for the subsequent nest.
duration since successful fledging until female initiates new nest (days)	Wf	3	
female body weight (g) during breeding season	BdyWt	6.3	"In summer in Riverside Co., CA, postbreeding adult body mass (measured at about 2 h after roosting time) was 6.3 g ± 0.6 SD (range 5.2–6.8 g, n = 6 males and females pooled)" as quoted in Webster (1999).

diet composition during breeding season	I		"Taylor (1971) analyzed contents of 41 adult and 42 immature Verdin stomachs from birds collected Feb–Oct in Maricopa Co., AZ. Data were not distinguished by season, nor was relative mass determined. Arthropods found from 7 orders. Lepidopterans (principally geometer [Geometridae] and gelechiid [Gelechiidae] moths) made up 21.1% of individual arthropods from adults and 44.6% of immature stomach contents. Homopterans (mainly aphids) were 28.5% of individuals identified from adult stomach contents and 22.3% of individuals found in stomachs from immature Verdins. Spiders made up large portion of diet of both adults (6.3% of individual arthropods) and immature Verdins (16.1%). Five nestlings collected in Apr and May had mainly caterpillars and spiders in their stomachs" as quoted in Webster (1999). During breeding season, assume diet of both adults and juveniles is 100% invertebrates.
mean number of nest attempts/ female/ season			"Second broods common; about half of Verdin pairs attempt 2 nesting cycles per season (Taylor 1971, Austin 1977)" as quoted from Webster (1999). "As many as 4 breeding attempts per season" as quote in Webster (1999).
mean number of successful broods/ female/ season			
mean number of fledglings/ successful nest	fpsn	3.0	Table 5 in Austin (1977) presents the "mean no. of fledglings per nest," but also contains enough information on the total number of nests and percentage of nests with fledged young to calculate the number of fledglings per successful nest. Weighted mean = 3.0 fledglings/successful nest.
mean number of fledglings/ female/ season (ARS)	ARS	3.42	Based on the review of data from multiple studies shown in Table 7 of Austin (1977), he calculated an average of 1.71 fledglings/adult or 3.42 fledglings/female.

Note: Although Austin (1977) indicates that "about half of Verdin pairs attempt 2 nesting cycles per season," meaning that there is a high probability of quitting breeding after a successful brood, in MCnest all females continue to initiate new nest attempts after success or failure until the end of the egg laying period. Consequently basic version of MCnest will overestimate the number of successful broods per season.

Reference List

Alisauskas, R. T., and C. D. Ankney. 1992. The cost of egg laying and its relationship to nutrient reserves in waterfowl. In: Ecology and Management of Breeding Waterfowl. B. Batt (Ed.), University of Minnesota Press. Pp 30-61.

Austin, G. T. 1977. Production and survival of the Verdin. Wilson Bull. 89:572-582.

Hensley, M. M. 1959. Notes on the nesting of selected species of birds of the Sonoran Desert. Wilson Bull. 71:86-92.

Taylor, W. K. 1967. Breeding biology and ecology of the Verdin, Auriparus flaviceps (Sundevall). Phd Thesis. Arizona State Univ. Tempe.

Taylor, W. K. 1971. A breeding biology study of the Verdin, Auriparus flaviceps (Sundevall) in Arizona. Am. Midl. Nat. 85:289-328.

Webster, Marcus D. 1999. Verdin (Auriparus flaviceps), The Birds of North America Online (A. Poole, Ed.). Ithaca: Cornell Lab of Ornithology; Retrieved from the Birds of North America Online: http://bna.birds.cornell.edu.bnaproxy.birds.cornell.edu/bna/species/470.

Carolina wren (*Thryothorus ludovicianus*) | **Four-letter Alpha Code: CARW**

Species life-history parameters	Model Code	Typical value	Rationale
daily mortality rate during laying & incubation	m1	0.043	"Of 118 nests, 31 (26%) produced at least 1 fledgling (TMH)" as quoted from Haggerty and Morton (1995). Using an apparent nest success rate of 26%, the daily nest mortality rate over 31 d (i.e., 13+14+5-1) is 0.043.
daily mortality rate during nestling-rearing	m2	0.043	
date of first egg of first nest (dd-mmm)	T1	1-Apr	"Egg date ranges: Illinois, 18 Apr–28 Jul (Bohlen 1989); Kentucky, 22 Mar–2 Aug (Mengel 1965); Florida, 1 Apr–24 Jun ($n = 44$; Bent 1948); Georgia, 5 Apr–3 Jul ($n = 33$; Bent 1948); Pennsylvania, 8 Apr–22 Jul ($n = 8$; Bent 1948); Texas, 13 Mar–9 Jul ($n = 39$; Bent 1948); nw. Alabama, 26 Mar–13 Aug, with 70% of eggs laid May–Jul ($n = 105$ clutches; TMH)" as quoted from Haggerty and Morton (1995). Based on Figure 3 in Haggerty and Morton (1995) assume egglaying from April 1 to August 3.
date of first egg of last nest (dd-mmm)	Tlast	3-Aug	
length of rapid follicle growth period (RFG) for each egg (days)	rfg	3	Assume rfg period similar to house wren of 3 d.
mean clutch size	clutch	5	"In a 2-yr study in nw. Alabama using nest boxes, modal clutch size 4 (mean 4.3, 0.7 SD, 3–6, $n = 88$; TMH). Of 255 nests in Western Foundation of Vertebrate Zoology collection, mode of 5 (mean 4.8, 0.6 SD, 3–7)" as quoted from Haggerty and Morton (1995). Usually 5 with range 4 to 8 (Ehrlich et al. 1988). Assume typical clutch size of 5.
mean intra-egg laying interval (days)	eli	1	"Eggs are laid on successive days and apparently within 1–2 h of sunrise (Nice and Thomas 1948, Rosene 1954)" as quoted from Haggerty and Morton (1995).
egg on which female typically begins incubation–penultimate (1) or last (0)	penult	1	"Incubation during day begins with penultimate or last egg of clutch, but female may spend night in nest after first egg is laid (Nice and Thomas 1948, Laskey 1950, TMH)" as quoted from Haggerty and Morton (1995). Assume incubation typically starts with penultimate egg.
duration from start of incubation to hatch (days)	I	14	"In nw. Alabama, 14.8 d (12–16, $n = 16$; TMH). Others report similar time period (Nice and Thomas 1948, Laskey 1948, 1950, Rosene 1954)" as quoted from Haggerty and Morton (1995). Incubation of 12-14 d (Ehrlich et al. 1988). Assume 14 d typical incubation period.
duration from hatch to fledging of nestlings (days)	N	13	"In nw. Alabama, young leave when 12.2 d old (10–16, $n = 5$; TMH), 13 d at a Tennessee nest (Laskey 1948), and 13–14 in Arkansas (Nice and Thomas 1948)" as quoted from Haggerty and Morton (1995). Assume typical nestling period is 13 d.
duration since nest failure due to other reasons until female initiates new nest (days)	We	9	"Renesting attempts are made if nests fail (Abbott 1884, James and Neal 1986, TMH). Interval between failed nests and renest: 9.3 d (5.6 SD, $n = 25$; TMH)" as quoted from Haggerty and Morton (1995).

duration since successful fledging until female initiates new nest (days)	Wf	11	"Fledglings stay together after leaving nest; no division of labor observed among parents. Both parents feed fledglings initially, but if female begins incubating subsequent clutch, male provides most of the care (TMH). In nw. Alabama, interval between fledging of young and laying of first egg of renest: 11 d (3.9 SD, $n = 12$; TMH)" as quoted from Haggerty and Morton (1995).
female body weight (g) during breeding season	BdyWt	18.6	"From nw. Alabama throughout year. Male 21.5 g (1.9 SD, $n = 26$); female 18.6 g (1.1 SD, $n = 26$) (a significant difference, $p = 0.0001$) (TMH)" as quoted from Haggerty and Morton (1995). Mean of 21.0 ± 1.15 (n=15)(Dunning 1984). Assume 18.6 g is typical female weight.
diet composition during breeding season		I	Primarily consume insects and spiders, with small amount of weed seeds. Based on Martin et al. (1951) in spring diet is 98% invertebrate and 2% seeds. Assume adults typically consume 98% invertebrates and 2 % seeds, while juveniles consume 100% invertebrates.
mean number of nest attempts/ female/ season		3.7	"In nw. Alabama, 3.7 nesting attempts were made per season (0.9 SD, 2–5, $n = 17$ pairs; TMH)" as quoted from Haggerty and Morton (1995).
mean number of successful broods/ female/ season		0.84	"In nw. Alabama, 31 females raised an average of 0.84 (0.86 SD) broods to nest-leaving per season. Eleven females (35%) raised 1 brood to nest-leaving, 6 (19%) raised 2 broods, and 1 (3%) raised 3 broods (TMH)" as quoted from Haggerty and Morton (1995).
mean number of fledglings/ successful nest	fpsn	3.8	"Number of fledglings/nest: 1.2 (1.8 SD, $n = 98$). Of 118 nests, 31 (26%) produced at least 1 fledgling (TMH)" as quoted from Haggerty and Morton (1995). If we interpret this to mean that a total of 118 fledglings (i.e., 98*1.2) were produced from 31 successful nests, the mean number of fledglings per successful nest is 3.8.
mean number of fledglings/ female/ season (ARS)	ARS	2.8	"Average annual reproductive success: 2.8 young reared to nest-leaving/female/season ($n = 40$)" as quoted from Haggerty and Morton (1995).

Reference List

Abbott, C. C. 1884. The Carolina Wren; a year of its life. *Am. Nat.* 18:21-25.

Bent, A. C. 1948. Life histories of North American nuthatches, wrens, thrashers, and their allies. *U.S. Natl. Mus. Bull.* 195.

Bohlen, H. D. 1989. The birds of Illinois. Indiana Univ. Press, Indianapolis, IN.

Dunning, J. B., Jr. 1984. Body weights of 686 species of North American birds. Western Bird Banding Association Monograph No. 1. 38 pp.

Ehrlich, P. R., D. S. Dobkin, and D. Wheye. 1988. The Birder's Handbook: A Field Guide to the Natural History of North American Birds. Simon & Schuster, Inc., New York. 785 pp.

Haggerty, T. M. and E. S. Morton. 1995. Carolina Wren (Thryothorus ludovicianus), The Birds of North America Online (A. Poole, Ed.). Ithaca: Cornell Lab of Ornithology; Retrieved from the Birds of North America Online: http://bna.birds.cornell.edu/bnaproxy.birds.cornell.edu/bna/species/188.

James, D. A. and J. C. Neal. 1986. Arkansas birds: their distribution and abundance. Univ. of Arkansas Press, Fayetteville.

Laskey, A. R. 1948. Some nesting data on the Carolina Wren at Nashville, Tennessee. *Bird-Banding* 19:101-121.

Laskey, A. R. 1950. A courting Carolina Wren building over nestlings. *Bird-Banding* 21:1-6.

Mengel, R. M. 1965. The birds of Kentucky. *Ornithol. Monogr.* 3.

Nice, M. M. and R. H. Thomas. 1948. A nesting of the Carolina Wren. *Wilson Bull.* 60:139-158.

Rosene, W. 1954. Nesting of the Carolina Wren. *Ala. Birdlife* 2:23-25.

Species Life-History Profiles – 12 December 2013

House wren (*Troglodytes aedon*)

Four-letter Alpha Code: HOWR

Species life-history parameters	Model Code	Typical value	Rationale
daily mortality rate during laying & incubation	m1	0.012	"Some data available on nest success of birds breeding under natural conditions—i.e., using natural cavities in relatively undisturbed habitats. In Wyoming, ≥1 nestling fledged from 62 (63%) of 99 natural nest sites observed from prelaying onward (Johnson and Kermott 1994), and from 35
daily mortality rate during nestling-rearing	m2	0.012	(70%) of 50 such nests in n. Arizona (Martin and Li 1992). In latter study, additional nests were found after laying began. Using Mayfield methods, daily nest mortality estimated at 0.010 and nest success at 75.5% (*n* = 115 nests). Also using Mayfield methods, Purcell et al. (1997) estimated daily nest mortality at 0.0126 and nest success at 63% (*n* = 47 nests) in e-central California (all nest failures attributed to predation)" as quoted from Johnson (1998). Assume that typical daily nest failure rate is 0.012.
date of first egg of first nest (dd-mmm)	T1	15-May	"Mean date of first egg in central Illinois: 6 May ± 4.5 d SD (range 29 Apr–14 May; *n* = 14; C. F. Thompson pers. comm.); majority of clutches are started 5–15 d after first clutch of season is started—usually between 11 and 21 May (see Fig. 1 in Drilling and Thompson 1991). Breeding begins earlier in California, with first eggs appearing in Apr at many low-elevation sites (e.g., Purcell et al. 1997). Breeding is later at higher latitudes and altitudes. Mean date of first egg in e.-central New York: 20 May ± 5 d SD (range 11–24 May, *n* = 5 yr; T. Alworth pers. comm.), in n.-central Wyoming mountain foothills (elevation 1,310 m): 19 May ± 3 d SD (range 14–23 May, *n* = 9 yr)" as quoted from Johnson (1998). Based on Figure 3 in Johnson (1998), assume typical date for T1 is May 15.
date of first egg of last nest (dd-mmm)	Tlast	15-Jul	"In central Illinois, where almost all females attempt second broods, production of most second clutches begins 50–60 d after first egg of season is laid—usually between about 24 Jun and 8 Jul (see Fig. 1 in Drilling and Thompson 1991). Mean date for laying of last egg in central Illinois: 4 Aug ± 5.1 d SD (range 28 Jul–13 Aug, *n* = 14 yr)" as quoted from Johnson (1998). Based on Figure 3 in Johnson (1998), assume typical date for Tlast is July 15.
length of rapid follicle growth period (RFG) for each egg (days)	rfg	3	Based on allometric equation (RFG = 2.852*Egg mass ^0.31) from Alisauskas and Ankney (1992), using and egg mass of 1.45 g, assume rfg = 3 d.
mean clutch size	clutch	6	"Reported mean clutch sizes include: nw. Ohio—6.0 ± 0.8 SD (*n* = 94 clutches, 2 yr; Robinson and Rotenberry 1991); central Illinois—6.4 (*n* = 3,781 clutches, 14 yr; C. F. Thompson pers. comm.); central Alberta—6.5 ± 1.1 SD (*n* = 46 clutches, 1 yr; Quinn and Holroyd 1992); s.-central British Columbia—6.4 ± SD (*n* = 40 clutches, 2 yr; Elliott et al. 1994); n.-central Wyoming—6.4 ± 1.1 SD (*n* = 254 clutches, 2 yr; LSJ)" as quoted from Johnson (1998).
mean intra-egg laying interval (days)	eli	1	"Usually lays 1 egg/d on successive days" as quoted from Johnson (1998).

92

Description	Symbol	Value	Notes
egg on which female typically begins incubation—penultimate (1) or last (0)	penult	1	"Substantial individual variation in nocturnal attentiveness during egg-laying, but in general, most females apply some heat to lone egg on first night, then gradually increase both time spent applying heat to eggs and intensity of heat each night thereafter until beginning "full" incubation through the night with laying of antepenultimate or penultimate egg (Kendeigh 1952: 84)" as quoted from Johnson (1998).
duration from start of incubation to hatch (days)	I	13	"In Wyoming, eggs begin hatching 12.6 d ± 1.0 SD (range 9–16, $n = 225$; LSJ) after day last egg is laid. Incubation period shortens as season progresses; e.g., in Illinois, incubation period for clutches completed May–mid-Jun: 12.7 d ± 0.9 SD ($n = 113$) versus 12.2 d ± 0.9 SD ($n = 66$) for clutches completed thereafter (Baltz and Thompson 1988; see also Milinkovich 1993: 27)" as quoted from Johnson (1998).
duration from hatch to fledging of nestlings (days)	N	16	"Young fledge on NSD 16, 17, or 18 (i.e., when first-hatched nestlings 15–17 d old). Usually all young leave nests within a few hours of one another" as quoted from Johnson (1998).
duration since nest failure due to other reasons until female initiates new nest (days)	We	8	No specific data found on duration of We, but assume renesting after failure requires building a new nest, which takes the female 3 to 14 d (Johnson 1998). Assume that a typical duration for We is 8 d.
duration since successful fledging until female initiates new nest (days)	Wf	8	"In n. Wyoming, mean interclutch interval (number of days between laying of last egg of first clutch and laying of first egg of second clutch) for individual females is 36.6 d ± 4.4 SD ($n = 26$ females)" as quoted from Johnson (1998). If incubation and nestling-rearing phases are a total of 29 d, the duration of Wf is approximately 8 d.
female body weight (g) during breeding season	BdyWt	12.0	"Mean mass of males captured in early incubation and late nestling stages of breeding in an Iowa population: 10.9 g ($n = 10$) and 10.7 g ($n = 12$), respectively (not significantly different; Freed 1981). In same population, female mass changed significantly during breeding cycle, averaging 12.0 g ($n = 42$) during incubation and 11.0 g ($n = 48$) and 10.5 g ($n = 29$) during the first and last halves of nestling stage, respectively" as quoted from Johnson (1998). Assume 12 g is a typical weight for females during most of the breeding season.
diet composition during breeding season		I	Diet of both adults and juveniles is primarily terrestrial invertebrates (Johnson 1998).
mean number of nest attempts/ female/ season			
mean number of successful broods/ female/ season			
mean number of fledglings/ successful nest	fpsn	5.5	No specific data found on the number of fledglings/successful nest, but Kendeigh (1942) reports that 92% of 5816 eggs laid in successful nests produced fledglings. If the mean clutch size is 6, the mean number of fledglings /successful nest is estimated to be 5.5.
mean number of fledglings/ female/ season (ARS)	ARS		

Reference List

Alisauskas, R. T., and C. D. Ankney. 1992. The cost of egg laying and its relationship to nutrient reserves in waterfowl. In: Ecology and Management of Breeding Waterfowl. B. Batt (Ed.), University of Minnesota Press. Pp 30-61.

Baltz, M. E. and C. F. Thompson. 1988. Successful incubation of experimentally enlarged clutches by House Wrens. *Wilson Bull.* 100:70-79.

Drilling, N. E. and C. F. Thompson. 1991. Mate switching in multibrooded House Wrens. *Auk* 108:60-70.

Elliott, J. E., P. A. Martin, T. W. Arnold, and P. H. Sinclair. 1994. Organochlorines and, reproductive success of birds in orchard and non-orchard areas of central British Columbia, Canada, 1990-1991. *Arch. Environ. Contam. Toxicol.* 26:435-443.

Freed, L. A. 1981. Loss of mass in breeding wrens: stress or adaptation? *Ecology* 62:1179-1186.

Johnson, L. S. and L. H. Kermott. 1994. Nesting success of cavity-nesting birds using natural tree cavities. *J. Field Ornithol.* 65:36-51.

Kendeigh, S. C. 1942. Analysis of losses in the nesting of birds. *J. Wildl. Manage.* 6:19-16.

Kendeigh, S. C. 1952. Parental care and its evolution in birds. *Illinois Biol. Monogr.* 18:1-356.

Martin, T. E. and P. Li. 1992. Life history traits of open- vs. cavity-nesting birds. *Ecology* 73:579-592.

Milinkovich, D. J. 1993. The sources of variation in the reproductive characters of House Wrens (*Troglodytes aedon*) breeding at two elevations in Colorado. Phd Thesis. Florida State Univ., Tallahassee.

Purcell, K. L., J. Verner, and L. W. Oring. 1997. A comparison of the breeding ecologies of four bird species nesting in boxes and tree cavities. *Auk* 114:646-656.

Quinn, M. S. and G. L. Holroyd. 1992. Asynchronous polygyny in the House Wren (*Troglodytes aedon*). *Auk* 109:192-195.

Robinson, K. D. and J. T. Rotenberry. 1991. Clutch size and reproductive success of House Wrens rearing natural and manipulated broods. *Auk* 108:277-284.

Blue-gray gnatcatcher (*Polioptila caerulea*)			Four-letter Alpha Code: BGGN
Species life-history parameters	Model Code	Typical value	Rationale
daily mortality rate during laying & incubation	m1	0.05	In IL, Kershner et al. (2001) calculated daily nest survival rates of 0.94 and 0.89 for first and second nests, respectively (overall rate of 0.93± 0.008 (n=93); equivalent to 11% overall nest success). In CA, young fledged from 24.4% of 42 nests found during construction phase (Root 1969), which for a 31 d nest period is equivalent to a daily nest survival rate of 0.955. In VT, 42% of nests found mid-incubation or earlier fledged young (Ellison 1991), which is equivalent to a daily nest survival rate of 0.972. Assume a typical daily success rate of 0.95 or daily failure rate of 0.05.
daily mortality rate during nestling-rearing	m2	0.05	
date of first egg of first nest (dd-mmm)	T1	May-1	Clutch initiation in Vermont (n=24) 16 May to 26 June, with 79% of initiations 16 May to 5 June (Ellison 1991). Egg dates in Ontario (n=22) 19 May to 1 July, with 50% from 28 May to 10 June (Peck and James 1987). In CA, eggs laid from early April thru mid-July (Fig 3; Root 1969). Nice (1932) found nests in late construction in mid-April in OK and in mid-May in OH.
date of first egg of last nest (dd-mmm)	Tlast	Jun-30	
length of rapid follicle growth period (RFG) for each egg (days)	rfg	3	RFG for passerines typically 3-4 d. Given the very low body weight, assume 3 d based on allometric equation for all birds from Alisauskas and Ankney (1992).
mean clutch size	clutch	5	"*P. c. caerula.* Of 182 clutches, a mean of 4.5 eggs/clutch. Frequency of 3-egg clutches: 13; 4-egg: 60; 5-egg: 107: 6-egg: 2. *P. c. amoenissima.* (160 clutches), mean of 4.35; 3-egg: 8; 4-egg: 88; 5-egg: 64." (quoted from Ellison 1992). Assume 5 as most common clutch size.
mean intra-egg laying interval (days)	eli	1	"One egg laid each day in the morning (Fehon 1955)." (quoted from Ellison 1992).
egg on which female typically begins incubation—penultimate (1) or last (0)	penult	0	"Incubation commences after last egg is laid (Root 1969)." (quoted from Ellison 1992).
duration from start of incubation to hatch (days)	I	13	"Mean 13 d, range 11-15 d (Weston 1949, Fehon 1955, Root 1969, Ellison 1991)." (quoted from Ellison 1992).
duration from hatch to fledging of nestlings (days)	N	13	"Departure from nest may occur 10-15 d after hatching: in sw. Vermont mean was 13 d (n=10; Ellison 1991)." (quoted from Ellison 1992).
duration since nest failure due to other reasons until female initiates new nest (days)	We	5	Figure 3 in Root (1969) indicates that nest building begins almost immediately after the failure of a nest attempt. Assume 5 days as the typical period for We.
duration since successful fledging until female initiates new nest (days)	Wf	5	"In c. CA (three pairs; Root 1969), building of second nest 5 d or less after two of the nests produced fledglings, within 10 d at a third. In sw. VT (Ellison 1991), 1988, second nests began immediately after first nests produce young." (quoted from Ellison 1992). Assume 5 days as the typical period for Wf.

95

female body weight (g) during breeding season	BdyWt	6.0	In PA, adult females in April 5.8 g ± 0.41 (n=42), in May 6.4 g ± 0.87 (n=27), and in August 6.0 g ± 0.40 (n=15) (Clench and Leberman 1977). In CA, breeding adults 5.7 g (n=13; range 5.4 – 6.0 g) (Root 1969). Assume mean female weight of 6.0 g.
diet composition during breeding season		I	Adult and juveniles consume 100% invertebrates – mostly small insects and spiders gleaned from broad-leaved foliage of trees and large shrubs (Ellison 1992)
mean number of nest attempts/ female/ season		1.9	"Overall, 42 females built 82 nests; an average of 1.9 ± 1.2 nests built/female (2.6 ± 1.2 nest/female that renested). The maximum number of nesting attempts documented for a female was 7" (Kershner et al 2001).
mean number of successful broods/ female/ season		0.61	"Most pairs are single brooded.....Seventeen of 23 females (74%) studied in VT raised at least one brood to fledgling stage (Ellison 1991). Only 3 (13%) succeeded in raising fledglings from 2 nests. 58% (7/12) of females studied in CA produced fledged young from first nests and 4 (33%) successfully fledged young from second broods (Root 1969)" (quoted from Ellison 1992). In IL, 15 of 42 pairs fledged at least one brood, with one of those successfully fledging a second brood (Kershner et al 2001). From these 3 studies a total of 47 successful broods from 77 pairs – 0.61 successful broods/female.
mean number of fledglings/ successful nest	fpsn	3	In IL, 3.4 ± 0.07 young/successful nest (n=10)(Kershner et al 2001). In VT, 2.7 young/successful attempt (n=18) (Ellison 1991). Assume average of 3 young/successful nest.
mean number of fledglings/ female/ season (ARS)	ARS	1.83	Based on values above: 0.61 * 3 = 1.83 fledglings/female/season.

Reference list

Alisauskas, R. T., and C. D. Ankney. 1992. The cost of egg laying and its relationship to nutrient reserves in waterfowl. Pp. 30-61 in: Ecology and Management of Breeding Waterfowl. (B.D. J.Batt, A. D. Afton, M. G. Anderson, C. D. Ankney, D. H. Johnson, J. A. Kadlec, and G. L. Krapu, Eds.). University of Minnesota Press, Minneapolis.

Clench, M. H. and R. C. Leberman. 1977. Weights of 151 species Pennsylvania birds analyzed by month, age, and sex. *Bull. Carnegie Mus. Nat. Hist.* No. 5.

Ellison, W. G. 1991. The mechanism and ecology of range expansion by the Blue-gray Gnatcatcher. Master's Thesis. Univ. Connecticut, Storrs.

Ellison, W. G. 1992. Blue-gray Gnatcatcher. In: The Birds of North America, No. 23. (A. Poole, P. Stettenheim, and F. Gill, Eds.). Philadelphia: The Academy of Natural Sciences; Washington, DC: The American Ornithologists' Union.

Fehon, J. H. 1955. Life-history of the Blue-gray Gnatcatcher (*Polioptila caerulea caerulea*). Phd Thesis. Florida State Univ., Tallahassee.

Kershner, E. L., E. K. Bollinger, and M. N. Helton. 2001. Nest-site selection and renesting in the Blue-gray Gnatcatcher (*Polioptila caerulea*). Amer. Midl. Natur. 146(2):404-413.

Nice, M. M. 1932. Observations on the nesting of the Blue-gray Gnatcatcher. *Condor* 34:18-22.

Peck, G. K. and R. D. James. 1987. Breeding birds of Ontario: nidiology and distribution, Vol. 2. Passerines. Life Sci. Misc. Publ., Roy. Ont Mus. Toronto.

Root, R. B. 1969. The behavior and reproductive success of the Blue-gray Gnatcatcher. *Condor* 71:16-31.

Weston, F. M. 1949. Blue-gray Gnatcatcher. Pages 345-355 *in* Life histories of North American thrushes, kinglets, and their allies. (A. C. Bent, Ed.) *U. S. Natl. Mus. Bull.* No. 196.

Eastern bluebird (*Sialia sialis*)	Four-letter Alpha Code: EABL		
Species life-history parameters	Model Code	Typical value	Rationale
daily mortality rate during laying & incubation	m1	0.0081	Based on Radunzel et al (1997) analysis of data on all nest types, 380 of 2644 nests (14.4%) failed during the egg phase. Assuming 19 d egg phase (14+5), daily success rate is 0.9919 so daily nest mortality rate is 0.0081
daily mortality rate during nestling-rearing	m2	0.0038	Based on Radunzel et al (1997) analysis of data on all nest types, 184 of 2644 nests (7.0%) failed during the nestling phase. Assuming 19 d nestling phase, daily success rate is 0.9962 so daily nest mortality rate is 0.0038
date of first egg of first nest (dd-mmm)	T1	1-Apr	Based on distribution of first egg dates for FL, TN, PA, and SC in Figure 6 of Gowaty and Plissner (1998), egg production typically begins about April 1 and ends mid July.
date of first egg of last nest (dd-mmm)	Tlast	15-Jul	
length of rapid follicle growth period (RFG) for each egg (days)	rfg	4	Based on allometric equation (RFG = $2.852*$Egg mass $^{0.31}$) from Alisauskas and Ankney (1992), using and egg mass of 3.6 g, assume rfg = 4 d.
mean clutch size	clutch	5	"Three to 6 or 7 eggs; in S. Carolina, modal clutch size 5 in spring, 4 in summer. Little variation in clutch size with latitude" (Gowaty and Plissner 1998).
mean intra-egg laying interval (days)	eli	1	Lays 1 egg per day (Gowaty and Plissner 1998).
egg on which female typically begins incubation–penultimate (1) or last (0)	penult	0	Incubation usually begins on the day the last egg is laid (Gowaty and Plissner 1998).
duration from start of incubation to hatch (days)	I	14	Incubation typically lasts 14 d, but can vary from 11-19 d (Gowaty and Plissner 1998).
duration from hatch to fledging of nestlings (days)	N	19	In MI mean age at nest departure from undisturbed nests is 18.8 d ± 1.47 SD (n=184); 19.4 and 18.6 d for spring and summer broods, respectively (Pinkowski 1975). In GA 17.6 d ± 1.8 SD (n=290) and in SC 17.6 d ± 1.2 SD (n=1298) (Gowaty and Plissner 1998).
duration since nest failure due to other reasons until female initiates new nest (days)	We	11	Little data on We. Many females that lose a nest prior to fledging leave the area. Based on a regression presented in Figure 3 of Pinkowski (1977), females with no surviving juveniles would begin egglaying in a new nest in 10.6 d, but it is not known if females losing nests early in laying or incubation would start sooner. Assume 11 d is a typical waiting period after unsuccessful nest.
duration since successful fledging until female initiates new nest (days)	Wf	20	"Adult birds who reared a spring brood renested in summer after an interval of 5 to 41 days (x = 19.50 ± 8.05 days, n=24) between the fledging of the spring brood and the onset of laying" (Pinkowski 1977). "Interval between fledging 1 brood and laying of next can be as long as several weeks. Typically, female time to renesting is about 2 wk after successful fledging" (Gowaty and Plissner 1998). Assume typical period of 20 d.

Species Life-History Profiles – 12 December 2013

female body weight (g) during breeding season	BdyWt	30	Based on Figure 7 of Gowaty and Plissner (1998) approximate female weight is 30 g.
diet composition during breeding season	I		Adults: mostly invertebrates (i.e., lepidoptera larvae, beetles, grasshoppers, crickets, spiders) with some fruit (most in fall and winter, less in spring). According to Martin et al. (1951) in April and May 7% fruit and 93% invertebrates. Juveniles: almost all invertebrates, so assume 100%.
mean number of nest attempts/ female/ season			
mean number of successful broods/ female/ season			
mean number of fledglings/ successful nest	fpsn	3.63	Mean number of fledglings/successful nest in MI is 3.73 ± 1.13 SD (n=299 nests) (Pinkowski 1977) and in SC 3.61 ± 1.16 SD (n=1520) (Gowaty and Plissner 1998). Weighted mean = 3.63 fledglings/successful nest.
mean number of fledglings/ female/ season (ARS)	ARS	5.0	Assuming equal proportions of second year (4.3 young/female/year) and after second year (5.7 young/female/year) birds, the population average productivity would be 5.0 young/female/year (Pinkowski 1979). Ricklefs and Bloom (1977) model estimate is 7.39 young/female/year.

Reference list

Alisauskas, R. T., and C. D. Ankney. 1992. The cost of egg laying and its relationship to nutrient reserves in waterfowl. In: *Ecology and Management of Breeding Waterfowl*. B. Batt (Ed.), University of Minnesota Press. Pp 30-61.

Gowaty, P. A. and J. H. Plissner. 1998. Eastern Bluebird (*Sialia sialis*), In: The Birds of North America Online (A. Poole, Ed.). Ithaca: Cornell Lab of Ornithology; Retrieved from the Birds of North America Online: http://bna.birds.cornell.edu.bnaproxy.birds.cornell.edu/bna/species/381

Martin, A. C., H. S. Zim, and A. L. Nelson. 1951. *American wildlife and plants: A guide to wildlife food habits*. Dover Publications, Inc., New York. 500 pp.

Pinkowski, B. C. 1975. Growth and development of eastern bluebirds. *Bird-Banding* 46:273-28.

Pinkowski, B. C. 1977. Breeding adaptations in the eastern bluebird. *Condor* 79:289-302.

Pinkowski, B. C. 1979. Annual productivity and its measurement in a multi-brooded passerine, the eastern bluebird. *Auk* 96:562-572.

Radunzel, L. A., D. M. Muschitz, V. M. Bauldry, and P. Arcese. 1997. A long-term study of the breeding success of eastern bluebirds by year and cavity type. *J. Field Ornithol.* 68:7-18.

Ricklefs, R. E., and G. Bloom. 1977. Components of avian breeding productivity. *The Auk* 94: 86-96.

Wood thrush (*Hylocichla mustelina*)

Four-letter Alpha Code: WOTH

Species life-history parameters	Model Code	Typical value	Rationale
daily mortality rate during laying & incubation	m1	0.046/0.030	Nest failure rates vary due to degree of forest fragmentation and cowbird parasitism. In fragmented forests, Donovan et al. (1995) reported daily nest mortality rates of 0.046 and 0.040 in MO and WI/MN, respectively, while at 15 sites in IN the rate was 0.035 (Fauth 2000) and at 3 sites in IL the mean was 0.061 (Brawn and Robinson 1996). *Assume the mean of these 4 sites (i.e., 0.046) is typical of fragmented forests.* In larger contiguous forests, Donovan et al. (1995) reported daily nest mortality rates of 0.031 and 0.018 in MO and WI/MN, respectively, while in Great Smokey NP the rate was 0.0415 (Simons et al. 2000). *Assume the mean of these 3 sites (i.e., 0.03) is typical of contiguous forests.*
daily mortality rate during nestling-rearing	m2	0.046/0.030	
date of first egg of first nest (dd-mmm)	T1	10-May	Among-year ($n = 19$) mean for first egg of year at UDW: 10 May ± 3.4 d (5–16 May) (Evans et al. 2011).
date of first egg of last nest (dd-mmm)	Tlast	21-Jul	Among-year mean date for latest clutch initiation: 21 Jul ± 7.0 ($n = 17$ yr; 12 Jul–1 Aug) (Evans et al. 2011).
length of rapid follicle growth period (RFG) for each egg (days)	rfg	5	Based on allometric equation (RFG = $2.852*$Egg mass$^{0.31}$) from Alisauskas and Ankney (1992), using and egg mass of 4.8 g, assume rfg = 5 d.
mean clutch size	clutch	3	In DE (UDW): "first nest: 3.7 ±0.6 (SD) ($n = 150$; 1–5); second: 3.0 ±0.6 ($n = 116$, 1–4); third: 2.8 ±0.4 ($n = 48$; 2–3); fourth: 2.6 ±0.5 ($n = 10$, 2–3). In DE, early season mean clutch size was 3.31 ±1.01, late season clutch size was lower at 2.56 ±0.77 (Brown and Roth 2002). May–Jun mean clutch size in Smoky Mtns., 1992–1995—was 3.5 (T. Simons and G. Farnsworth pers. comm.)" as quoted from Evans et al. (2011). Assume typical clutch size of 3.
mean intra-egg laying interval (days)	eli	1	Eggs laid 1/d in succession, beginning usually 2–3 d after construction has ended (Evans et al. 2011).
egg on which female typically begins incubation–penultimate (1) or last (0)	penult	1	Because incubation usually begins with penultimate egg, first hatching often occurs twelfth day after clutch completed (Evans et al. 2011).
duration from start of incubation to hatch (days)	I	13	For 59 unparasitized clutches of 2–4 eggs with complete hatches at UDW (1990–1993), mean time from last egg laid to last hatched 12.7 d ± 0.7 (11–14; RRR) (Evans et al. 2011).
duration from hatch to fledging of nestlings (days)	N	14	Nestlings fledge at 12–15 d (Evans et al. 2011).
duration since nest failure due to other reasons until female initiates new nest (days)	We	7	"Laying was resumed more quickly-on the seventh day, and not later than the eighth day-after the destruction of two nests by storms on the third and eighth days of incubation, respectively" as quoted from Brackbill 1958. Assume We of 7 d.
duration since successful fledging until female initiates new nest (days)	Wf	10	"At four second-brood nests the first eggs were laid 9, 9, 11 and 12 days, respectively, after the young had left the previous nests" as quoted from Brackbill 1958. Assume Wf of 10 d.

Parameter	Code	Value	Notes
female body weight (g) during breeding season	BdyWt	52.5	From Table 1 in Evans et al. (2011): In DE, mean female wt during breeding season is 52.5 g ± 4.4 (n=955); in NH, mean of 50.2 g (n=7) and in PA, mean of 50.1 g (n=20).
diet composition during breeding season		O	Adults: primarily soil invertebrates and some fruit. In the summer months 65% invertebrates and 35% fruit (Martin et al. 1951).
mean number of nest attempts/ female/ season			In IN, mean of 3.1 nest attempts/yr for color-marked population (Fauth 2000).
mean number of successful broods/ female/ season			"Typically double-brooded; rarely three broods/yr (Brackbill 1958, RRR). Rates of double brooding vary in s. Ontario; 74% of females double brooded and double brooding may be as high as 87% (Friesen et al. 2000). In nw. Pennsylvania only 50% of females attempted a second brood; double brooding highly constrained by high nest predation (Gow 2009)" as quoted from Evans et al. (2011).
mean number of fledglings/ successful nest	fpsn	2.0/3.0	The number of fledglings varies due to degree of forest fragmentation and cowbird parasitism. In fragmented forests in MO and WI/MN the mean number of fledglings/successful nest was 2.14 and 2.42, respectively (Donovan et al. 1995), while the mean was 1.4 at 15 sites in IN (Fauth 2000). *Assume the mean of these 3 sites (i.e., 2.0) is typical of fragmented forests.* In DE (UDW) mean of 2.6 ± 1.0 fledglings/successful nest (n=287) from 1974-1994 (Evans et al. 2011). In Great Smokies Mt. NP mean of 3.31 fledglings/successful nest (n=153) from 1992 to 1997 (Simons et al. 2000). In contiguous forests in MO and WI/MN the mean was 3.02 and 3.0, respectively (Donovan et al. 1995). *Assume the mean of these 4 sites (i.e., 3.0) is a typical values for contiguous forests.*
mean number of fledglings/ female/ season (ARS)	ARS	1.8/3.4	ARS varies due to degree of forest fragmentation and cowbird parasitism. In fragmented forests, Donovan et al. (1995) reported ARS of 1.7 and 2.24 in MO and WI/MN, respectively, while at 15 sites in IN the rate was 1.8 (Fauth 2000). Weinberg and Roth (1998) reported the mean number of fledglings/female/yr in small forest fragments in DE for two years was 1.5. *Assume the mean of these 4 sites (i.e., 1.8) is typical of fragmented forests.* Donovan et al. (1995) reported ARS of 3.5 and 4.8 in MO and WI/MN, respectively, while in larger contiguous forests, while in Great Smokey NP from 1992 to 1997 the rate was 2.76 (Simons et al. 2000). In DE (UDW) "among-year mean in 1974–1994 (except 1979) was 2.6 ± 0.7 fledglings/female/yr (n = 20; range 1.7–3.8; Roth and Johnson 1993, RRR)" as quoted from Evans et al. (2011). *Assume the mean of these 4 sites (i.e., 3.4) is typical of contiguous forests.*

Reference List

Alisauskas, R. T., and C. D. Ankney. 1992. The cost of egg laying and its relationship to nutrient reserves in waterfowl. In: *Ecology and Management of Breeding Waterfowl*. B. Batt (Ed.), University of Minnesota Press. Pp 30-61.

Brackbill, H. 1958. Nesting behavior of the Wood Thrush. *Wilson Bull.* 70:70-89.

Brawn, J. D. and S. K. Robinson. 1996. Source-sink population dynamics may complicate the interpretation of long-term census data. *Ecology* 77:3-12.

Brown, W. P. and R. R. Roth. 2002. Temporal patterns of fitness and survival in the Wood Thrush. *Ecology* 83(4):958-969.

Donovan, T. M., F. T. Thompson, and J. Faaborg. 1995. Reproductive success of migratory birds in habitat sources and sinks. *Conserv. Biol.* 9:1380-1395.

Evans, Melissa, Elizabeth Gow, R. R. Roth, M. S. Johnson and T. J. Underwood. 2011. Wood Thrush (*Hylocichla mustelina*), The Birds of North America Online (A. Poole, Ed.). Ithaca: Cornell Lab of Ornithology; Retrieved from the Birds of North America Online: http://bna.birds.cornell.edu/bna/species/246

Fauth, P. T. 2000. Reproductive success of Wood Thrushes in forest fragments in northern Indiana. *Auk* 117(1):194-204.

Gow, A. E. 2009. Carry-over effects in Wood Thrush (*Hylocichla mustelina*): linking reproduction to moult. M.Sc. Thesis. York University, Toronto, ON.

Martin, A. C., H. S. Zim, and A. L. Nelson. 1951. American wildlife and plants: A guide to wildlife food habits. Dover Publications, Inc., New York. 500 pp.

Roth, R. R. and R. K. Johnson. 1993. Long-term dynamics of a Wood Thrush population breeding in a forest fragment. *Auk* 110:37-48.

Simons, T. R., G. L. Farnsworth, and S. A. Shriner. 2000. Evaluating Great Smoky Mountains National Park as a population source for the Wood Thrush. *Conserv. Biol.* 14(4):1133-1144.

Weinberg, H. J. and R. R. Roth. 1998. Forest area and habitat quality for nesting Wood Thrushes. *Auk* 115(4):879-889.

American robin (*Turdus migratorius*)			Four-letter Alpha Code: AMRO
Species life-history parameters	Model Code	Typical value	Rationale
daily mortality rate during laying & incubation	m1	0.025	Mayfield estimate of daily nest mortality rate for 257 nests in PNW = 0.05 and overall nest success of 0.26 (0.95^26=0.26) (Sallabanks and James 1999); Based on mean of 10 studies in table below with mean daily success of 0.975; Daily nest mortality rate in KS of 0.0221 (Ricklefs and Bloom 1977)
daily mortality rate during nestling-rearing	m2	0.023	Based on mean of 10 studies in table below with mean daily success of 0.977; Daily nest mortality rate in KS of 0.0221 (Ricklefs and Bloom 1977).
date of first egg of first nest (dd-mmm)	T1	12-Apr	Egg laying dates: April 12 – July 22 in Wisconsin (Young 1955); April 6 – July 24 in New York (Howell 1942); May 10 to July 6 in northern Maine (Knupp et al. 1977); May 5-July 20 in Pacific Northwest w/ peak egg laying (50% of nests) June 5 to July 4 (Sallabanks and James 1999);
date of first egg of last nest (dd-mmm)	Tlast	22-Jul	Mean date of first egg from 11 studies in Table 2 of Knupp et al 1977 is April 10; Use the Wisconsin egg dates as "typical values" (April 12-July 22) (Julian dates: 203-102= 101 d); Breeding season length of 3.14 months or 94 d (Ricklefs andBloom 1977).
length of rapid follicle growth period (RFG) for each egg (days)	rfg	4	Based on allometric equation from Alisauskas and Ankney (1992), assume rfg = 4 d.
mean clutch size	clutch	3	Usually 3 or 4 egg (rarely 5) (Sallabanks and James 1999); 3.6 in KS (Ricklefs and Bloom 1977).
mean intra-egg laying interval (days)	eli	1	One egg laid daily until completion (Sallabanks and James 1999).
egg on which female typically begins incubation–penultimate (1) or last (0)	penult	0	"Incubation may begin after second egg is laid (Schantz 1939), perhaps by application of only partial heat (Kendeigh 1952)" (Sallabanks and James 1999). "If the weather is mild, setting does not commence until the complete clutch is laid" (Howe 1898). Assume, in most cases actual incubation begins with last egg.
duration from start of incubation to hatch (days)	I	13	Approximately 13 d (range 12 to 14) after laying last egg (Sallabanks and James 1999).
duration from hatch to fledging of nestlings (days)	N	13	Fledge about day 13 (range 9-16; Howell 1942).
duration since nest failure due to other reasons until female initiates new nest (days)	We	7	No specific info yet on time to renesting after failure; so assume same duration as after success; Estimate of 10 d in KS from Ricklefs and Bloom (1977).

duration since successful fledging until female initiates new nest (days)	Wf	7	"Approximately 7 d separate fledging of young from first nest and laying of first egg of second clutch (Howell 1942)" (Sallabanks and James 1999); Estimate of 10 d in KS from Ricklefs and Bloom (1977); "The next clutch is usually started about 40 days after the first egg of the year, but females often start the second nest, including laying the eggs, before the first group of young is independent. Sometimes the overlap is extensive, with the second clutch begun before the first nestlings are out of the nest. When this happens, the male cares for the first nestlings." (www.hww.ca).
female body weight (g) during breeding season	BdyWt	77.3	77.3 g ± 0.36 SD (n=401) for males and females in PA throughout year (Clench and Leberman 1978); 75.0 g ± 2.4 SE (n=2) in OR during breeding (Sallabanks and James 1999).
diet composition during breeding season		O	USEPA (1993) Wildlife Exposure Factors Handbook summarizes Wheelwright (1986) analysis of robin diet (fruit vs invertebrates) by three regions (eastern, central, and western US) and four seasons. The percentage of invertebrates in spring was 93, 92, and 83% across regions and in summer was 32, 24, and 37%. Average across spring and summer seasons and 3 regions is 72% invertebrates and 28% fruits. Juveniles fed 70% invertebrates and 30% plant material, which is assumed to be fruit (Howell 1942 as cited in Sallabanks and James 1999).
mean number of nest attempts/ female/ season			
mean number of successful broods/ female/ season		2	Regularly rear 2 broods/season, sometimes 3, especially in southern portion of range (Sallabanks and James 1999).
mean number of fledglings/ successful nest	fpsn	2.8	In Maine, mean number of young/successful nest = 2.5 ± 0.15 SE (n=38; Knupp et al 1977); North American Nest Record Cards for the northeastern states had average of 3.3, 3.0, and 3.0 fledglings/successful nest in 1966, 1967, & 1968 (Johnson et al 1976); also 2.86 and 2.4 reported by Young 1955 and Howell 1942; mean of six studies above = 2.8 fledglings/successful nest.
mean number of fledglings/ female/ season (ARS)	ARS	5	Estimates of # young/female/season: 3.9 (Howell 1942), 5 (Farner 1945), 5.6 (Young 1949, 1955, 1956) though these are criticized for overestimation of % successful nests (Sallabanks and James 1999); 5.2 fledglings/female/season (Ricklefs and Bloom 1977). Assume approximately 5 fledglings/female/year.

Reference List

Alisauskas, R. T., and C. D. Ankney. 1992. The cost of egg laying and its relationship to nutrient reserves in waterfowl. In: *Ecology and Management of Breeding Waterfowl.* B. Batt (Ed.), University of Minnesota Press. Pp 30-61.

Farner, D. S. 1945. Age groups and longevity in the American robin. *The Auk* 57, no. 1: 56-74.

Howe, R. H. Jr. 1898. Breeding habits of the American robin (*Merula migratoria*) in eastern Massachusetts. *The Auk* 15: 162-67.

Howell, J. C. 1942. Notes on the nesting habits of the American Robin (*Turdus migratorius* L.). *Amer. Midl. Natural.* 28:529-603.

Johnson, E. V., G. L. Mack, and D. Q. Thompson. 1976. The effects of orchard pesticide applications on breeding robins. *Wilson Bull.* 88:16-35.

Knupp, D. M., R. B. Owen Jr., and J. B. Dimond. 1977. Reproductive biology of American robins in northern Maine. *The Auk* 94: 80-85.

Ricklefs, R. E., and G. Bloom. 1977. Components of avian breeding productivity. *The Auk* 94: 86-96.

Sallabanks, R. and F. C. James. 1999. American Robin (*Turdus migratorius*). In The Birds of North America, No. 462 (A. Poole and F. Gill, Eds.). The Birds of North America, Inc., Philadelphia, PA.

Schantz, W. E. 1939. A detailed study of a family of robins. *Wilson Bull.* 51:157-169.

Young, H. 1949. A comparative study of nesting birds in a 5 acre park. *Wilson Bull.* 61:36-47.

Young, H. 1955. Breeding behavior and nesting of the eastern robin. *Amer. Midl. Natural.* 53: 329-352.

Young, H. 1956. Territorial activities of the American Robin *Turdus migratorius*. *Ibis* 98:448-452.

Northern mockingbird (*Mimus polyglottos*) — Four-letter Alpha Code: NOMO

Species life-history parameters	Model Code	Typical value	Rationale
daily mortality rate during laying & incubation	m1	0.031	Apparent nest success in PA & MD averaged 36.4%; in FL averaged 39.4%; and in IL averaged 61%. Nest success varied considerable among years in PA & MD (from 6.7% to 80%), but varied less in FL. Late nests are more successful than early in IL (Derrickson and Breitwisch 1992). Assume 40% overall nest success is typical. Over a 29 d nesting period (i.e., 12+13+4), the daily nest survival rate would be 0.969, and daily nest mortality rate is 0.031.
daily mortality rate during nestling-rearing	m2	0.031	
date of first egg of first nest (dd-mmm)	T1	1-Apr	In FL and NC, nest building starts as early as late February, although March is more common (Howell 1932, RB). Complete nests were found as early as February 19th in n. FL, with eggs laid on March 2nd (GAL, JUM). Northern populations begin building nests 3-5 wk later, beginning in mid-April and continuing through mid- to late August (Sprunt 1964). In se. PA and MD, nesting can begin late March (mid-April more common), terminating during August (KCD) (Farnsworth et al. 2011). Assume typical start and end dates of April 1 and August 1.
date of first egg of last nest (dd-mmm)	Tlast	1-Aug	
length of rapid follicle growth period (RFG) for each egg (days)	rfg	4	Based on allometric equation (RFG = 2.852*Egg mass ^0.31) from Alisauskas and Ankney (1992), using and egg mass of 4.2 g, assume rfg = 4 d.
mean clutch size	clutch	4	Mean clutch size in FL 3.5 eggs (n=156, range=2-5, Derrickson and Breitwisch 1992), in LA 3.7 eggs (n=266, range 2-6, Means & Goertz 1983), in TN 3.9 eggs (n=212, range 3-5, Laskey 1962), in IL 3.8 eggs (n=52, range 3-6, Graber et al 1970), and in PA & MD 3.6 eggs (n=88, range 2-5, Derrickson and Breitwisch 1992). Assume typical nest has clutch size of 4.
mean intra-egg laying interval (days)	eli	1	One egg per day (Derrickson and Breitwisch 1992).
egg on which female typically begins incubation–penultimate (1) or last (0)	penult	0	"Incubation is sporadic until the clutch is complete, becoming more constant with the penultimate egg" as quoted from Derrickson and Breitwisch (1992). Since eggs generally hatch within 24 hr period, assume incubation effectively begins with last egg.
duration from start of incubation to hatch (days)	I	13	12 to 13 d in FL, NC and PA and 13 d in LA and IL (Derrickson and Breitwisch 1992).
duration from hatch to fledging of nestlings (days)	N	12	Most nestlings depart the nest (usually in early morning) on the 12th day after hatching, although may leave as early as the 10th day (especially if disturbed) and as late as the 15th day (Derrickson and Breitwisch 1992).
duration since nest failure due to other reasons until female initiates new nest (days)	We	7	Information in Logan 1983 and Derrickson 1989 indicate nest building can start almost immediately to a few days after a nest failure and takes about 4 d, with egg laying starting shortly after completion. Assume female initiates laying after failure in 7 d.
duration since successful fledging until female initiates new nest (days)	Wf	7	"After nest departure, both parents feed young for one to several days, and then the male nearly or completely stops providing and begins to construct the foundation for the next nest. After a few days, the male resumes providing, and the female stops feeding the young as she finishes the nest, lays and incubates the eggs" as quoted in Derrickson and Breitwisch (1992). Assume female initiates laying in 7 d.

female body weight (g) during breeding season	BdyWt	46.7	Mean female body weights: 46.4 g (SD=3.9, n=44, FL), 47.2 (SD=3.1, n=27, NC), 46.5 (SD=4.5, N=11, DC),and 47.0 (SD=4.4, n=5, PA) from Derrickson and Breitwisch (1992). Weighted mean = 46.7 g.
diet composition during breeding season		O	Adults consume primarily invertebrates and fruit. During breeding season the proportion of invertebrates increases to nearly 85% (Derrickson and Breitwisch 1992). Assume breeding season diet of 85% invertebrates and 15% fruit. For approximately the first week nestlings are fed almost entirely invertebrates, but the proportion of fruit increases until reaching 20-30% of volume in older nestlings (Breitwisch et al. 1984). Assume younger nestlings are more vulnerable and that their diet consists of 100% invertebrates.
mean number of nest attempts/ female/ season			In se. Pennsylvania and Maryland, the average number of nesting attempts/female/year was 2.7 (n = 43, range = 1–7); In FL mean of 3.2 nesting attempts/female (n=68, range 2-6 attempts) (Derrickson and Breitwisch 1992).
mean number of successful broods/ female/ season		1.0	Based on data in Derrickson and Breitwisch (1992) in PA & MD 43 females had 43 successful broods (30% had 0, 42% had 1, 26% had 2 and 2% had 3). In FL, Zaias and Breitwisch (1989) had 27 pairs produce 47 successful broods (i.e., 1.74 successful broods/female) over two years.
mean number of fledglings/ successful nest	fpsn	2.8	Mean number of fledglings/successful nest in FL in two years was 2.7 (SD=0.9, n=22, range 1–4) and 2.4 (SD=1.0, n=25, range 1–4) (Zaias and Breitwisch 1989); in LA was 3.0 (SD1.0, n=101) (Means and Goertz 1983); and in PA and MD was 2.5 (SD=0.9, n=43nests, range 1–4) (Derrickson and Breitwisch 1992). The weighted mean of the 4 values is 2.8 fledglings/successful nest.
mean number of fledglings/ female/ season (ARS)	ARS	2.5	Mean number of fledglings/female/year is 2.5 (SD=2.0, n=43 females, range 0-8) in se PA and MD (Derrickson and Breitwisch 1992). Ricklefs and Bloom (1977) calculated ARS of 2.27.

Note: Although several references describe an egglaying period lasting approximately 4 months, Ricklefs and Bloom (1977) used a period of 1.92 months in AZ. The longer egglaying period used in MCnest results in a much higher number of nest attempts and successful nests than reported in the literature. This suggests that either the probability of quitting after success and failure is relatively high or that the reports of egglaying dates reflect extremes that poorly represent the overall population.

Reference list

Alisauskas, R. T., and C. D. Ankney. 1992. The cost of egg laying and its relationship to nutrient reserves in waterfowl. In: *Ecology and Management of Breeding Waterfowl*. B. Batt (Ed.), University of Minnesota Press. Pp 30-61.

Derrickson, K. C. 1989. Bigamy in Northern Mockingbirds: Circumventing female-female aggression. *Condor* 91:728-732.

Derrickson, K. C. and R. Breitwisch. 1992. Northern Mockingbird. In: The Birds of North America, No. 7 (A. Poole, P. Stettenheim, and F. Gill, Eds.). Philadelphia: The Academy of Natural Sciences; Washington, DC.: The American Ornithologists' Union.

Farnsworth, G., G. A. Londono, J. U. Martin, K. C. Derrickson and R. Breitwisch. 2011. Northern Mockingbird (*Mimus polyglottos*), The Birds of North America Online (A. Poole, Ed.). Ithaca: Cornell Lab of Ornithology; Retrieved from the Birds of North America Online: http://bna.birds.cornell.edu.bnaproxy.birds.cornell.edu/bna/species/007.

Howell, A. H. 1932. Florida bird life. Coward-McCann, New York.

Logan, C. A. 1983. Reproductively dependent song cyclicity in mated male mockingbirds (*Mimus polyglottos*). *Auk* 100:404-413.

Means, L. L. and J. W. Goertz. 1983. Nesting activities of Northern Mockingbirds in northern Louisiana. *Southwest. Nat.* 28:61-70.

Ricklefs, R. E., and G. Bloom. 1977. Components of avian breeding productivity. *The Auk* 94: 86-96.

Zaias, J. and R. Breitwisch. 1989. Intra-pair cooperation, fledgling care, and renesting by Northern Mockingbirds. *Ethology* 80:94-110.

Cedar waxwing (*Bombycilla cedrorum*)			Four-letter Alpha Code: CEDW
Species life-history parameters	Model Code	Typical value	Rationale
daily mortality rate during laying & incubation	m1	0.015	Weighted mean of nest success (% successful nests of total nests) for 4 studies (i.e., MI, OH, WI, Ontario) reported in Table 2 of Witmer et al. (1997) was 64%. Over a 31 d nest period (i.e. 16+12+4-1), the daily nest success rate is 0.985, so daily nest mortality rate is 0.015.
daily mortality rate during nestling-rearing	m2	0.015	Dates of first eggs in Witmer et al. (1997) focus on extreme dates and peak dates from three studies, but histograms of first egg dates in two of the studies (Leck and Cantor 1979, Putnam 1949) provide a means for determining typical dates for start and end of first eggs in new nests by cutting off the tails of distributions of dates involving very few birds. Assume typical dates June 8 to August 15.
date of first egg of first nest (dd-mmm)	T1	8-Jun	
date of first egg of last nest (dd-mmm)	Tlast	15-Aug	
length of rapid follicle growth period (RFG) for each egg (days)	rfg	4	Based on allometric equation from Alisauskas and Ankney (1992), assume rfg = 4 d based on 3.2 g egg.
mean clutch size	clutch	4	In OH, mean of 4.15 and range 2-5 (Putnam 1949). In Ontario, mean of 4.18 and range 2-6 (Mountjoy 1987). In northeast US and south Canada nest cards, mean 4.2 (Leck and Cantor 1979).
mean intra-egg laying interval (days)	eli	1	Eggs laid daily in the morning (Witmer et al. 1997).
egg on which female typically begins incubation–penultimate (1) or last (0)	penult	1	Proportion of time female spends incubating increases with each egg laid, so hatching is asynchronous (over 48 hr period), so assume using incubation start with penultimate egg is adequate description.
duration from start of incubation to hatch (days)	I	12	About 12 d from laying to hatching, range 11-13 d (Witmer et al. 1997).
duration from hatch to fledging of nestlings (days)	N	16	"Average nestling period is 15.5 d (range 14-18, n=21 young; Saunders 1911, Lea 1942)" as quoted from Witmer et al. (1997). Assume typical nestling period of 16 d.
duration since nest failure due to other reasons until female initiates new nest (days)	We	7	Little data on We. Unlike successful nests where females anticipate the end of first nest and begin preparing for egg production in the second, most nest failures would mean female would need to build new nest and physiologically prepare for RFG period. Assume this requires about 7 d.
duration since successful fledging until female initiates new nest (days)	Wf	5	According to Putnam (1949) there is considerable overlap of activity between first and second nest, with females reducing brooding activity during last week on first nest to resume courtship and nest building on second. "The laying of the first egg in the second nest varies from the day before fledging at the first nest to 3 days after" (Putnam 1949). However, the basic version of MCnest is not designed to allow this overlap of the nestling rearing phase with the RFG and egglaying phases, so Wf is set to 5 d.

Species Life-History Profiles – 12 December 2013

female body weight (g) during breeding season	BdyWt	34	Mean of 17 females during Jun thru Aug = 34.0 (Witmer et al. 1997).
diet composition during breeding season		F	Adults: Martin et al. (1951) indicate during the months of June thru August adults consume about 80% fruit and 20% insects. Juveniles are fed predominantly insects for the first day or two, but proportion of fruit increases with age until fledging when juveniles are fed almost entirely fruit (Witmer et al. 1997). Assume over the nestling period 80% fruit and 20% insects.
mean number of nest attempts/ female/ season			
mean number of successful broods/ female/ season		≤2	"In Ohio, mates often remain together to raise 2 broods per season (7 of 8 pairs; Putman 1949). In se. Ontario, less than one-third of pairs estimated to raise 2 broods (Mountjoy 1987)" as quoted in Witmer et al. (1997).
mean number of fledglings/ successful nest	fpsn	3.63	Weighted mean of fledglings/successful nests for 4 studies (i.e., MI, OH, WI, Ontario) reported in Table 2 of Witmer et al. (1997).
mean number of fledglings/ female/ season (ARS)	ARS		

Reference list

Alisauskas, R. T., and C. D. Ankney. 1992. The cost of egg laying and its relationship to nutrient reserves in waterfowl. In: *Ecology and Management of Breeding Waterfowl.* B. Batt (Ed.), University of Minnesota Press. Pp 30-61.

Lea, R. B. 1942. A study of the nesting habits of the Cedar Waxwing. *Wilson Bull.* 54:225-237.

Leck, C. F. and F. L. Cantor. 1979. Seasonality, clutch size, and hatching success in the Cedar Waxwing. *Auk* 96:196-198.

Martin, A. C., H. S. Zim, and A. L. Nelson. 1951. *American wildlife and plants: A guide to wildlife food habits.* Dover Publications, Inc., New York. 500 pp.

Mountjoy, D. J. 1987. Behavioural ecology of the Cedar Waxwing during the breeding season. Master's Thesis. Queen's Univ., Kingston, Ontario.

Putnam, L. S. 1949. The life history of the Cedar Waxwing. *Wilson Bull.* 61:141-182.

Saunders, A. A. 1911. A study of the nesting of the Cedar Waxwing. *Auk* 28:323-329.

Witmer, M. C., D. J. Mountjoy and L. Elliot. 1997. Cedar Waxwing (*Bombycilla cedrorum*), The Birds of North America Online (A. Poole, Ed.). Ithaca: Cornell Lab of Ornithology; Retrieved from the Birds of North America Online: http://bna.birds.cornell.edu.bnaproxy.birds.cornell.edu/bna/species/309.

Ovenbird (*Seiurus aurocapillus*)	Four-letter Alpha Code: OVEN		
Species life-history parameters	Model Code	Typical value	Rationale
daily mortality rate during laying & incubation	m1	0.050/0.034	Most of these studies analyze nest success data using the Mayfield method (Mayfield 1975) -- estimating daily survival and extrapolating nest success from this based on the number of days in the nesting cycle. In Missouri, nest success was 27% (n = 15) in forest fragments and 38% (n = 40) in contiguous forest (Donovan et al. 1995). In Wisconsin, 20% (n = 36) of nests were successful in forest fragments, with 42% (n = 51) successful in contiguous forest (Donovan et al. 1995). Also in Wisconsin, nest success was 44% (n = 42) for nests neighboring clearcuts and 69% (n = 47) for nests in the forest interior (Flaspohler et al. 2001a). In contiguous forest in Minnesota, nest success was 41% (n = 318, Manolis et al. 2002). In Tennessee in the Great Smoky Mountains National Park, nest success was 30% (n = 89, Podolsky et al. 2007). In extensive forest in n. New Hampshire, 47% of nests were successful (n = 98, King et al. 1996). In e. Massachusetts, 29% (n = 62) of nests in suburban forest fragments were successful. In Ontario, nests in small fragments had significantly lower success (15%, n = 74) than nests in large fragments (44%, n = 61) which had significantly lower success than nests in contiguous forest (70%, n = 30, Burke and Nol 2000)" as quoted from Porneluzi et al. (2011). Weighted mean of nest success is 30% and 44% in fragmented and contiguous forests, respectively. For a 24 d nest period (i.e., 4+12+8), daily nest mortality rates in fragmented and contiguous forests are 0.050 and 0.034, respectively.
daily mortality rate during nestling-rearing	m2	0.050/0.034	
date of first egg of first nest (dd-mmm)	T1	May-15	"Bent (1953) reported egg date ranges of May 15 to June 3 (n = 14) in Pennsylvania; May 16 to July 26 (n = 12) in Michigan; May 17 to June 14 (n = 66) in Massachusetts; and May 27 to June 21 (n = 17) in Quebec. In Tennessee, earliest nest initiation on 14 April, latest on 20 July (Podolsky et al. 2007; 110 nests over 3 years)" as quoted in Porneluzi et al. (2011). Based on Figure 3 in Porneluzi et al. (2011), core egg laying occurs from May 15 to June 30.
date of first egg of last nest (dd-mmm)	Tlast	Jun-30	
length of rapid follicle growth period (RFG) for each egg (days)	rfg	4	RFG for passerines typically 3-4 d.
mean clutch size	clutch	4	"Clutch size ranges from 3 to 6 (Bent 1953). Mean clutch size is 4.4 from 78 clutches (Western Foundation for Vertebrate Zoology). In Vermont, 22 clutches averaged 4 eggs each (Ellison 1985). In Missouri Ozarks with minimal rate (4%) of brood parasitism, mean of 48 nests was 4.2 + 0.1 (SE) (Porneluzi and Faaborg 1999). In Wisconsin, where parasitism was also low (2%) the mean was 4.9 + 0.1 (SE) for 39 nests near a clearcut edge and 4.3 + 0.1 for 40 nests in the interior of extensive forest (Flaspohler et al. 2001a, b). In studies with no brood parasitism, the mean size of 29 clutches was 4.2 + 0.1 (SE) in New Hampshire (King and Degraff 2002) and the mean of 89 clutches in Great Smoky Mountain National Park in Tennessee was 4.5 + 0.03 (SE; clutch size ranged from 3 to 6, Podolsky et al. 2007)" as quoted from Porneluzi et al. (2011). Assume typical clutch size of 4 eggs.

Description	Code	Value	Notes
mean intra-egg laying interval (days)	eli	1	One egg laid each day until clutch is complete.
egg on which female typically begins incubation–penultimate (1) or last (0)	penult	0	"Females begin incubating in the late afternoon or early evening on the day before the last egg is laid" as quoted from Porneluzi et al. (2011).
duration from start of incubation to hatch (days)	I	12	"In Michigan, mean, 12 d, 5.6 h; range, 11 d, 12 h to 14 d (26 eggs from 21 nests). Second clutches may require a shorter incubation period" as quoted from Porneluzi et al (2011).
duration from hatch to fledging of nestlings (days)	N	8	"In Michigan, young fledge at an average of 7 and 22.5 h (n = 57 birds from 16 nests; Hann 1937), and 8–10 d in Vermont (Ellison 1985). In North Carolina average nestling period was 7.6 ± 0.12 days and was inversely related to food abundance (Stodola et al. 2010)" as quoted from Porneluzi et al. (2011).
duration since nest failure due to other reasons until female initiates new nest (days)	We	5	"If the nest is destroyed during the early stages of incubation, the female will renest. Second nests can be completed in 4 d" as quoted from Porneluzi et al. (2011). Assume first egg in new nest typically in 5 d.
duration since successful fledging until female initiates new nest (days)	Wf	30	"As the other young depart, the brood is split between the parents and at this time young are coaxed to move about as adults offer food. Age of dispersal estimated by telemetry was 29.2 days after fledging (Vitz and Rodewald 2010)" as quoted from Porneluzi et al. (2011). Independent of adult care at approximately 30 d (Hann 1937). Assume any renesting after success occurs after fledglings are independent of adult.
female body weight (g) during breeding season	BdyWt	19.4	In PA, 19.4 ± 1.22 (range 14.0-28.8, n=181) (Dunning 1984).
diet composition during breeding season		I	"Major components of the adult diet include: Curculionidae, Coleoptera adults and larvae, Formicidae, Lepidoptera larvae, Diptera and Hemiptera adults (Stenger 1958, Holmes and Robinson 1988). Nestlings are fed primarily Carabidae, Lepidoptera larvae, and other larvae" as quoted from Porneluzi et al. (2011). Assume both adults and juveniles consume 100% invertebrates.
mean number of nest attempts/ female/ season			"Typically 1 clutch/pair/yr; occasionally 2 if the first nesting attempt fails (Bent 1953). In Michigan, Ovenbirds averaged 1.5 nests/pair (Hahn 1937)" as quoted from Porneluzi et al. (2011).
mean number of successful broods/ female/ season			"Hann (1937) reports only one record of a male Ovenbird raising a second brood after successfully raising the first. Podolsky et al. (2007) report three instances of double brooding in Tennessee. Porneluzi and Faaborg (1999) did not observe double brooding in their study of color-banded individuals in Missouri" as quoted from Porneluzi et al. (2011).
mean number of fledglings/ successful nest	fpsn	3.2/4.2	In MO, number of fledglings/successful nest was 3.16 and 4.3 in fragmented and contiguous forest, respectively, while in WI/MN 3.3 and 4.0 in fragmented and contiguous forest, respectively (Donovan et al. 1995). Assume mean of two locations as typical value: 3.2 and 4.2, respectively.

| mean number of fledglings/ female/ season (ARS) | ARS | 1.33/2.75 | "In Michigan, Ovenbirds averaged 2.9 fledglings/pair per breeding season; only 1.6 of these 2.9 fledglings survived to independence each season (Hahn 1937)" as quoted from Porneluzi et al. (2011). In MO, number of fledglings/female/year was 1.48 and 2.64 in fragmented and contiguous forest, respectively, while in WI/MN 1.18 and 2.7 in fragmented and contiguous forest, respectively (Donovan et al. (1995). Mean in fragmented forest at 2 locations (MO & WI/MN) is 1.33 and in contiguous forest at 3 locations (MI, MO & WI/MN) is 2.75 fledglings/female/year. |

Note: "Fledging occurs when young are able to walk or hop away from the nest. Flight is developed after fledging (see Fledgling stage, growth). At fledging, young have achieved approximately 73% of the adult mass" as quoted from Porneluzi et al. (2011).

Reference List

Bent, A. C. 1953. Life histories of North American wood warblers. *U.S. Natl. Mus. Bull.* No. 203.

Burke, D. M. and E. Nol. 2000. Landscape and fragment size effects on reproductive success of forest-breeding birds in Ontario. *Ecol. Appl.* 10(6):1749-1761.

Donovan, T. M., F. R. Thompson, J. Faaborg, and J. R. Probst. 1995. Reproductive success of migratory birds in habitat sources and sinks. *Cons. Biol.* 9(6):1380-1395.

Dunning, J. B., Jr. 1984. Body weights of 686 species of North American birds. Western Bird Banding Association Monograph No. 1. 38 pp.

Ellison, W. 1985. Ovenbird. Pages 314 *in* The atlas of breeding birds of Vermont. (Laughlin, S. B. and D. P. Kibbe, Eds.) Univ. Press of New England, Hanover, NH.

Flaspohler, D. J., S. A. Temple, and R. N. Rosenfield. 2001a. Effects of forest edges on Ovenbird demography in a managed forest landscape. *Cons. Biol.* 15(1):173-183.

Flaspohler, D. J., S. A. Temple, and R. N. Rosenfield. 2001b. Species-specific edge effects on nest success and breeding bird density in a forested landscape. *Ecol. Appl.* 11(1):32-46.

Hann, H. W. 1937. Life history of the Ovenbird in scouthern Michigan. *Wilson Bull.* 44:146-235.

Holmes, R. T. and S. K. Robinson. 1988. Spatial patterns, foraging tactics, and diets of ground-foraging birds in a northern hardwoods forest. *Wilson Bull.* 100:377-394.

King, D. I. and R. M. DeGraaf. 2002. The effect of forest roads on the reproductive success of forest-dwelling passerine birds. *For. Sci.* 48(2):391-396.

King, D. I., C. R. Griffin, and R. M. Degraaf. 1996. Effects of clearcutting on habitat use and reproductive success of the ovenbird in forested landscapes. *Cons. Biol.* 10(5):1380-1386.

Mayfield, H. F. 1975. Suggestions for calculating nest success. *Wilson Bull.* 87:456-466.

Podolsky, A. L., T. R. Simons, and J. A. Collazo. 2007. Modeling population growth of the Ovenbird (*Seiurus aurocapilla*) in the southern Appalachians. *Auk* 124(4):1359-1372.

Porneluzi, P. A. and J. Faaborg. 1999. Season-long fecundity, survival, and viability of Ovenbirds in fragmented and unfragmented landscapes. *Cons. Biol.* 13(5):1151-1161.

Porneluzi, P., M. A. Van Horn, and T.M. Donovan. 2011. Ovenbird (*Seiurus aurocapilla*), The Birds of North America Online (A. Poole, Ed.). Ithaca: Cornell Lab of Ornithology; Retrieved from the Birds of North America Online: http://bna.birds.cornell.edu.bnaproxy.birds.cornell.edu/bna/species/088.

Stenger, J. 1958. Food habits and available food of Ovenbirds in relation to territory size. *Auk* 75:335-346.

Stodola, K. W., D. A. Buehler, D. H. Kim, K. E. Franzreb, and E. T. Linder. 2010. Biotic and abiotic factors governing nestling-period length in the Ovenbird (*Seiurus aurocapilla*). *Auk* 127(1):204-211.

Common yellowthroat (*Geothlypis trichas*) — Four-letter Alpha Code: COYE

Species life-history parameters	Model Code	Typical value	Rationale
daily mortality rate during laying & incubation	m1	0.0263	Following data from 2-yr study of shrubland and grassland habitats of Minnesota and Michigan: In Minnesota, 27/43 eggs hatched (62.8%) in 12 nests; of these, 15 young fledged (34.9% of eggs laid, 55.6% of eggs that hatched); nest success (fledged at least one young) was 33.3% (4/12 nests). In Michigan, 73/109 eggs hatched (67.0%) in 38 nests; of these, 64 young fledged (58.7% of eggs laid, 87.7% of eggs that hatched); nest success was 55.3% (21/38 nests; Hofslund 1959)" as quoted from Guzy and Ritchison (1999). Weighted mean of MN & MI data = 50% nest success (25/50 nests). Based on a 50% apparent nest success and 26 d nest period (i.e., 10+12+4), the daily nest mortality rate is 0.0263.
daily mortality rate during nestling-rearing	m2	0.0263	
date of first egg of first nest (dd-mmm)	T1	8-May	"Bent (1953) reported egg dates of 30 May–22 Jul for Arizona, 4 Apr–10 Jul for California, 23 May–27 Jun for Massachusetts, and 4 Jun–26 Jun for Nova Scotia. Other dates: British Columbia, 30 Apr to 7 Jul (R. W. Campbell pers. comm.); Ontario, 19 May–29 Jul (n = 135 nests; Peck 1987); S. Dakota, 9 Jun, male feeding fledgling 16 Aug (S. Dakota Ornithol. Union 1991); Arkansas, 5 May–24 Jun, recently fledged young being fed 16 Aug (James and Neal 1986); earliest laying date in n. Minnesota, 4 Jun; s. Michigan, 19 May (Hofslund 1959); Wisconsin, 26 May to 7 Jul for (Robbins 1991); Illinois, 7 May–9 Jul (Bull 1974); Massachusetts, 24 May to 17 Jun (Veit and Petersen 1993); Maryland and District of Columbia, 4 May–4 Aug, young in the nest 17 May–22 Aug, peak mid-Jun (Robbins and Blom 1996); Florida, 10 Apr to 24 Jul (Stevenson and Anderson 1994" as quoted from Guzy and Ritchison (1999). Mean dates for the above ranges are May 8 to July 11, which are very similar to those in IL and NY.
date of first egg of last nest (dd-mmm)	Tlast	11-Jul	
length of rapid follicle growth period (RFG) for each egg (days)	rfg	3	Based on allometric equation from Alisauskas and Ankney (1992), assume rfg = 3 d for 1.66 g egg.
mean clutch size	clutch	4	"One to 6 eggs, usually 4" as quoted from Guzy and Ritchison (1999).
mean intra-egg laying interval (days)	eli	1	"Usually one egg laid/d until clutch is completed" as quoted from Guzy and Ritchison (1999).
egg on which female typically begins incubation—penultimate (1) or last (0)	penult	0	"Incubation by female only; begins with laying of last egg of clutch" as quoted from Guzy and Ritchison (1999).
duration from start of incubation to hatch (days)	I	12	About 12 d (Stewart 1953, Hofslund 1959).
duration from hatch to fledging of nestlings (days)	N	10	10d (Ehrlich et al.1988). By day 8, vocalizing, able to leave nest (Stewart 1953, Hofslund 1959).

Parameter	Abbr.	Value	Notes
duration since nest failure due to other reasons until female initiates new nest (days)	We	5	"May desert nests if eggs are removed or damaged by cowbirds, or if cowbirds lay eggs in nest before yellowthroats do (Hofslund 1957). A second or even a third nest may be built on top of a parasitized nest (Butler 1898, cited in Bent 1953)" as quoted from Guzy and Ritchison (1999). No specific quantitative data on We. Assume after nest failure, renesting occurs rapidly with a typical duration for We of 5 d.
duration since successful fledging until female initiates new nest (days)	Wf	7	"The following information based on Hofslund 1959: In populations that are double-brooded, female apparently feeds only the first brood's fledglings during the first few days, after which the male cares for the fledglings" as quoted from Guzy and Ritchison (1999). No specific quantitative data on duration of Wf. Assume female care for fledglings for a few days and renests rapidly with duration of Wf of 7 d.
female body weight (g) during breeding season	BdyWt	9.9	"$G.\ t.\ trichas$: males, mean 10.3 g ± 0.66 SD (range 7.6–15.5, $n = 965$), females 9.9 g ± 0.78 SD (range 7.6–15.3, $n = 644$); $G.\ t.\ occidentalis$: males, mean 10.0 g (9.5–10.7, $n = 14$); $G.\ t.\ scirpicola$: males, mean 9.2 g (8.2–10.0, $n = 11$; Dunning 1993). No data on seasonal variation" as quoted from Guzy and Ritchison (1999).
diet composition during breeding season		I	"In Arizona, stomach contents of 10 birds contained (in approximate decreasing frequency) spiders, hemipterans, dipterans (flies), coleopterans (beetles), ants and termites, various larvae, hymenopterans (bees and wasps), grasshoppers, and homopterans (Rosenberg et al. 1982). Analysis of 11 stomachs (Indiana): 22 case-bearing caterpillars, 5 other larvae, 6 small dragonflies, 3 moths, 3 flies, 3 small hymenopterans (bees and wasps), 3 beetles, 3 spiders, 2 small grasshoppers, 1 leafhopper, 2 hemipterans, and 2 insect eggs (Butler 1898, cited in Bent 1953). One stomach from a bird collected in Massachusetts contained beetles, flies, and small seeds (Townsend 1905, cited in Bent 1953). Food brought to a nest (Iowa) over the entire nesting period (1,694 observations) consisted of 376 unidentified insects, 347 moths, 290 larvae of various kinds, 280 spiders, 116 mayflies, 61 flies, 92 unrecognized material, 20 caterpillars, 54 damselflies, 13 beetles, 13 crysalids, 11 butterflies, 10 seeds, 3 caddisflies, and 6 grasshoppers (Shaver 1918)" as quoted from Guzy and Ritchison (1999). Assume both adults and juveniles consume 100% invertebrates.
mean number of nest attempts/ female/ season			
mean number of successful broods/ female/ season			"Double-brooded in s. Michigan (Hofslund 1959) and s. Texas (Klicka 1994); single-brooded in n. Minnesota (Hofslund 1959)" as quoted from Guzy and Ritchison (1999).
mean number of fledglings/ successful nest	fpsn	3.16	In MN & MI, total of 79 fledglings from 25 successful nest, so 3.16 fledglings/successful nest (from Guzy and Ritchison 1999).
mean number of fledglings/ female/ season (ARS)	ARS		

Reference List

Alisauskas, R. T., and C. D. Ankney. 1992. The cost of egg laying and its relationship to nutrient reserves in waterfowl. In: *Ecology and Management of Breeding Waterfowl*. B. Batt (Ed.), University of Minnesota Press. Pp 30-61.

Bent, A. C. 1953. Life histories of North American wood warblers. *U.S. Natl. Mus. Bull.* 203.

Bohlen, H. D. 1989. The birds of Illinois. Indiana Univ. Press, Bloomington.

Bull, J. 1974. Birds of New York State. Doubleday, Garden City, New York.

Butler, A. W. 1898. The birds of Indiana. Indiana Dep. Geol. Nat. Res. 22nd Ann. Rep.

Dunning, Jr., J. B. 1993. CRC handbook of avian masses. CRC Press, London.

Ehrlich, P. R., D. S. Dobkin, and D. Wheye. 1988. The Birder's Handbook: A Field Guide to the Natural History of North American Birds. Simon & Schuster, Inc., New York. 785 pp.

Guzy, M. J. and G. Ritchison. 1999. Common Yellowthroat (*Geothlypis trichas*), The Birds of North America Online (A. Poole, Ed.). Ithaca: Cornell Lab of Ornithology; Retrieved from the Birds of North America Online: http://bna.birds.cornell.edu/bnaproxy.birds.cornell.edu/bna/species/448.

Hofslund, P. B. 1957. Cowbird parasitism of the Northern Yellowthroat. *Auk* 74:42-48.

Hofslund, P. B. 1959. A life history study of the Yellowthroat, *Geothlypis trichas*. *Proc. Minn. Acad. Sci.* 27:144-174.

Imhof, T. A. 1976. Alabama birds. Univ. of Alabama Press, University.

Klicka, J. T. 1994. The biological and taxonomic status of the Brownsville Yellowthroat (*Geothlypis trichas insperata*). Master's Thesis. Univ. of Minnesota, Minneapolis.

Peck, G. K. 1987. Breeding birds of Ontario: nidiology and distribution. Royal Ontario Museum, Toronto.

Robbins, C. S. and E. A. T. Blom. 1996. Atlas of the breeding birds of Maryland and the District of Columbia. Univ. of Pittsburgh Press, Pittsburgh.

Robbins, Jr., S. D. 1991. Wisconsin birdlife. Univ. of Wisconsin Press, Madison.

Rosenberg, K. V., R. D. Ohmart, and B. W. Anderson. 1982. Community organization of riparian breeding birds: Response to an annual resource peak. *Auk* 99:260-274.

Shaver, N. E. 1918. A nest study of the Maryland Yellow-throat. *Univ. Iowa Stud. Nat. Hist.* 8:1-12.

South Dakota Ornithologists' Union. 1991. The birds of South Dakota. 2nd ed. S. Dakota Ornithol. Union, Aberdeen.

Stevenson, H. M. and B. H. Anderson. 1994. The birdlife of Florida. Univ. Press of Florida, Gainesville.

Stewart, R. E. 1953. A life history study of the Yellowthroat. *Wilson Bull.* 65:99-115.

Townsend, C. W. 1905. The birds of Essex County Massachusetts. Mem. *Nuttall Ornithol. Club* 3.

Veit, R. and W. Petersen. 1993. Birds of Massachusetts. Mass. Audubon Soc., Lincoln.

Yellow warbler (*Setophaga petechia*)			Four-letter Alpha Code: YWAR
Species life-history parameters	Model Code	Typical value	Rationale
daily mortality rate during laying & incubation	m1	0.0176	"Daily survival of nests (which measures the probability a nest survives 1 d without predation or other loss) in n. Manitoba during incubation 0.969 ± 0.008 SD ($n = 61$ nests) and with young in the nest 0.974 ± 0.012 SD ($n = 36$ nests; Briskie 1995); daily survival of nests prior to initiation of incubation 0.9469 ± 0.180 SD ($n = 208$ nests, 320 d); during incubation 0.9870 ± 0.035 SD ($n = 177$ nests, 1,852 d); during nestling period 0.9765 ± 0.064 SD ($n = 113$ nests, 639 d); and from first to last fledge 0.9924 ± 0.062 SD ($n = 92$ nests, 266 d; Hébert and Sealy 1993b)" as quoted in Lowther et al. (1999). Weighted mean daily nest success rates for incubation and nestling rearing phases are 0.9824 and 0.9759, respectively.
daily mortality rate during nestling-rearing	m2	0.0241	
date of first egg of first nest (dd-mmm)	T1	25-May	"Late May to mid-Jun. Egg dates: Texas records: 17 May–13 Jul (Oberholser 1974); Ontario nest records: 15 May–17 Jul ($n = 967$), most of these 2 Jun–15 Jun ($n = 483$; Peck and James 1987); s. Manitoba study: clutches initiated between 26 May–7 Jul, with 50% of clutches initiated within 2–8 d interval (10–11 Jun 1974, 3–8 Jun 1975, 28 May–4 Jun 1976; Goossen and Sealy 1982); for n. Manitoba, 14 Jun–10 Jul (Briskie 1995); for central Alberta: 30 May–21 Jun, most (115 of 127) 2–13 Jun (CC); for British Columbia: 10 May–16 Aug, most (250 of 455) 7–23 Jun (Campbell et al. 2001); for se. Alaska, 24 May–26 Jun ($n = 20$; Rogers 1994). One brood normally reared; second broods only rarely attempted (Goossen and Sealy 1982)" as quoted in Lowther et al. (1999). Assume that core egg laying typically from May 25 to June 15.
date of first egg of last nest (dd-mmm)	Tlast	15-Jun	
length of rapid follicle growth period (RFG) for each egg (days)	rfg	3	Pearson and Rohwer (1998) reported rfg periods of 3 days in related spp.: Hermit warbler (*Dendroica occidentalis*) and Townsend's warbler (*Dendroica townsendi*).
mean clutch size	clutch	4	"Mean clutch size shows latitudinal increase in size from about 2.5 eggs in the West Indies to 4.5 eggs in Canada (see Briskie 1995; Prather and Cruz 1995; also Schrantz 1943, Graber et al. 1983, Wiley 1985, Kessel 1989, Sealy 1992, Rogers 1994. *D. p. babad*: St. Lucia, mean 2.3 ± 0.61 SD ($n = 61$); *D. p. cruciana*: Puerto Rico, 3 studies, mean 2.3 ± 0.78 SD ($n = 61$); *D. p. gundlachi*: s. Florida, mean 2.5 ± 0.73 SD ($n = 16$); and *D. p. aestiva*: Illinois, 4.0 ($n = 33$); Iowa, 4.5 ($n = 41$); s. Manitoba, 4.5 ± 0.02 SD ($n = 1,005$); cf. *D. p. parkesi* n. Manitoba, 4.6 ± 0.51 SD ($n = 54$); cf. *D. p. rubiginosa* se. Alaska, 5.11 ± 0.57 SD ($n = 20$); *D. p. banksi* w. Alaska, 4.9 ± 0.7 SD (range 36, $n = 14$), mode 5 ($n = 10$)" as quoted in Lowther et al. (1999). Assume typical clutch size in continental U. S. is 4.
mean intra-egg laying interval (days)	eli	1	"Eggs laid 1/d, approximately 24 h apart (McMaster et al. 1999); occasionally days skipped in laying sequence (Smith 1943)" as quoted in Lowther et al. (1999).
egg on which female typically begins incubation—penultimate (1) or last (0)	penult	1	Incubation "begins before clutch is completed" as quoted in Lowther et al. (1999). Assume incubation starts with penultimate egg.

Parameter	Code	Value	Notes
duration from start of incubation to hatch (days)	I	11	"Measured as interval between last egg laid and last egg hatching: 11 d (n = 1; Bigglestone 1913); 11 d (range 11–12, n = 6; Schrantz 1943). For s. Manitoba, 11.3 d ± 0.47 SD (n = 22); for n. Manitoba, 11.7 d ± 0.81 SD (range 11–13 d, n = 15; Briskie 1995); for central Alberta, 10 d (n = 6), 11 d (n = 16), 12 d (n = 6) or 13 d (n = 1; CC); for Colorado, 10.5 d ± 1.2 SD (n = 16; Ortega 1998)" as quoted in Lowther et al. (1999). Assume 11 d is typical.
duration from hatch to fledging of nestlings (days)	N	8	"Young leave nest 8–10 d after hatching; in s. Manitoba, mean 8.2 d ± 0.80 SD (n = 12); in n. Manitoba, mean 8.5 d ± 0.64 SD (n = 14; Briskie 1995)" as quoted in Lowther et al. (1999). Assume 8 d is typical.
duration since nest failure due to other reasons until female initiates new nest (days)	We	5	No data on duration of We if nest is predated or destroyed by weather. Yellow warblers are known to build over a parasitized nest and start a new clutch, probably very rapidly after abandoning first clutch. Assume that if nest has to be rebuilt, the earliest a new clutch could start is 5 d.
duration since successful fledging until female initiates new nest (days)	Wf	17	"One brood normally reared; second broods only rarely attempted (Goossen and Sealy 1982). Young still with adults at 17 d and possibly 21 d after leaving nest (Smith 1943)" as quoted in Lowther et al. (1999). Although second clutch after success in rare, assume it would not occur until young from first clutch are independent of adults. Using an egg laying period from May 25 to June 15 means there is not sufficient time for renesting after success.
female body weight (g) during breeding season	BdyWt	9.6	"Mean mass, males: 10.0 g ± 0.6 SD (n = 178); females: 9.6 g ± 0.8 SD (n = 140), for period 21–27 May at Delta Marsh, Manitoba (Biermann and Sealy 1985). Extensive data taken 1976–1982 demonstrate seasonal dynamics of mass on breeding grounds (Biermann and Sealy 1985): Male mass generally constant for first 10 wk of breeding then increases for final 5 wk prior to migration (weekly means range from 9.9–10.4 g); female mass initially drops after arrival, increases during egg-laying and incubation, then drops and gradually returns to their arrival value after young become independent in Jul (weekly means range from 9.6–10.9 g)" as quoted in Lowther et al. (1999).
diet composition during breeding season	I	I	"Main Foods Taken: Insects and other arthropods; may take wild fruits occasionally (Stevenson and Anderson 1994)" as quoted in Lowther et al. (1999). Assume adults and juveniles consume 100% invertebrates.
mean number of nest attempts/ female/ season			
mean number of successful broods/ female/ season			
mean number of fledglings/ successful nest	fpsn	3.3	Goossen and Sealy (1982) reported mean of 3.3 fledglings per successful nest over 3 year study in Manitoba.
mean number of fledglings/ female/ season (ARS)	ARS		

Species Life-History Profiles – 12 December 2013

Reference List

Biermann, G. C. and S. G. Sealy. 1985. Seasonal dynamics of body mass of insectivorous passerines breeding on the forested dune ridge, Delta Marsh, Manitoba. *Can. J. Zool.* 63:1675-1682.

Bigglestone, H. C. 1913. A study of the nesting behavior of the Yellow Warbler (*Dendroica aestiva aestiva*). *Wilson Bull.* 25:49-67.

Briskie, J. V. 1995. Nesting biology of the Yellow Warbler at the northern limit of its range. *J. Field Ornithol.* 66:531-543.

Campbell, R. W., N. K. Dawe, I. Mctaggert-Cowan, J. M. Cooper, and G. W. Kaiser. 2001. The Birds of British Columbia. Vol. 4. R. Br. Columbia Mus. Victoria, British Columbia.

Goossen, J. P. and S. G. Sealy. 1982. Production of young in a dense nesting population of Yellow Warblers, *Dendroica petechia*, in Manitoba. *Can. Field-Nat.* 96:189-199.

Graber, J. W., R. R. Graber, and E. L. Kirk. 1983. Illinois birds: Wood warblers. Ill. Nat. Hist. Surv., *Biol. Notes* 118:1-144.

Hébert, P. N. and S. G. Sealy. 1993b. Hatching asynchrony in Yellow Warblers: A test of the nest-failure hypothesis. *Ornis Scand.* 24:10-14.

Kessel, B. 1989. Birds of the Seward Peninsula, Alaska: Their biogeography, seasonality, and natural history. Univ. of Alaska Press, Fairbanks.

Lowther, P. E., C. Celada, N. K. Klein, C. C. Rimmer and D. A. Spector. 1999. Yellow Warbler (Setophaga petechia), The Birds of North America Online (A. Poole, Ed.). Ithaca: Cornell Lab of Ornithology; Retrieved from the Birds of North America Online: http://bna.birds.cornell.edu.bnaproxy.birds.cornell.edu/bna/species/454.

Mcmaster, D. G., S. G. Sealey, S. A. Gill, and D. L. Neudorf. 1999. Timing of egg laying in Yellow Warblers. *Auk* 116:236-240.

Oberholser, H. C. 1974. The bird life of Texas. Univ. of Texas Press, Austin.

Ortega, C. 1998. Cowbirds and other brood parasites. Arizona Univ. Press, Tucson.

Peck, G. and R. James. 1987. Breeding birds of Ontario: nidiology and distribution. Vol. 2. Passerines. R. Ontario Mus. Life Sci. Misc. Publ. Toronto.

Prather, J. W. and A. Cruz. 1995. Breeding biology of Florida Prairie Warblers and Cuban Yellow Warblers. *Wilson Bull.* 107:475-484.

Rogers, C. M. 1994. Avian nest success, brood parasitism and edge-independent reproduction in an Alaskan wetland. *J. Field Ornithol.* 65:433-440.

Schrantz, F. G. 1943. Nest life of the Eastern Yellow Warbler. *Auk* 60:367-387.

Sealy, S. G. 1992. Removal of Yellow Warbler eggs in association with cowbird parasitism. *Condor* 94:40-54.

Smith, W. P. 1943. Some Yellow Warbler observations. *Bird-Banding* 14:57-63

Stevenson, H. M. and B. H. Anderson. 1994. The birdlife of Florida. Univ. Press of Florida, Gainesville.

Wiley, J. W. 1985. Shiny Cowbird parasitism in two avian communities in Puerto Rico. *Condor* 87:165-176.

Yellow-rumped warbler (*Setophaga coronata*)

Four-letter Alpha Code: YRWA

Species life-history parameters	Model Code	Typical value	Rationale
daily mortality rate during laying & incubation	m1	0.022	"Nesting success (probability of successfully fledging at least 1 young) varies geographically. At Mogollon Rim, AZ, 51.5% (T. Martin unpubl.). At Snake River, ID, 25.2% success (S. Garner and L. Garner unpubl.). In San Bernardino Mtns., CA, 37.5% (G. Geupel unpubl.). In Sierra Nevada, CA, 100% (K. Purcell unpubl.)." as quoted from Hunt and Flaspohler (1998). The mean of these four unpublished estimates is 54%. If 54% is a typical estimate for apparent nest success rate, the daily nest mortality rate over a 28 d nest period (i.e., 12+12+4) is 0.022.
daily mortality rate during nestling-rearing	m2	0.022	
date of first egg of first nest (dd-mmm)	T1	25-May	"Mean nest initiation date for 12 nests in n. Wisconsin was 4 Jun (range 25 May–9 Jul; DJF). Mean nest initiation date for 130 nests in Arizona was 30 May (range 4 Mar–3 Jul; T. Martin unpubl.). Egg dates come from variety of locations and sources and may reflect first or second broods or renests. 24 May–29 Jul in Ontario (Peck and James 1987); 13 May–11 Jun in Maine and New Hampshire (Forbush 1929, Knight 1908); 18 May–6 Aug in New York (DJF, Cornell Nest Records Program [CNRP] cards); 1 Jun–26 Jul in Michigan (Wood 1951, DJF, CNRP cards; $n = $ 75 cards). Initiation dates (first egg laid) for 12 nests in n. Wisconsin ranged from 23 May to 28 Jun; peak initiation in last week of May and first week of Jun. Mean 24 May in s. Arizona ($n = $ 130; Brandt 1951); 12 Jun–1 Jul in Colorado (Baily and Niedrach 1965); 9 Apr–27 Jun in Washington (Dawson 1909, Jewett et al. 1953); 21 Apr in Portland, OR (Carnegie Museum of Natural History nest cards, DJF); 16 Jun in Colorado above 2,300 m (Forbush 1929); and 7 Jun in Alaska (Gabrielson and Lincoln 1957)" as quoted from Hunt and Flaspohler (1998). Based on Figure 4 in Hunt and Flaspohler (1998), assume typical core egg laying dates range from May 25 to June 30.
date of first egg of last nest (dd-mmm)	Tlast	30-Jun	
length of rapid follicle growth period (RFG) for each egg (days)	rfg	3	Typical rfg period for passerines of 3-4 d.
mean clutch size	clutch	4	"Typically 4–5, sometimes 3 (Harrison 1975), occasionally 6 (Jewett et al. 1953, Peck and James 1987). Of 53 nests in Ontario: 5 one-egg nests, 4 two-egg nests, 10 three-egg nests, 24 four-egg nests, 8 five-egg nests, 2 six-egg nests (Peck and James 1987). Of 8 nests in n. Wisconsin, 5 four-egg nests, 3 five-egg nests (mean 4.38 ± 0.48 SD; DJF). Of 14 nests in Arizona, mean clutch size 3.86 ± 0.36 SD (range 3–4; T. Martin unpubl.)." as quoted from Hunt and Flaspohler (1998).
mean intra-egg laying interval (days)	eli	1	"1 egg laid/d until clutch is complete, although occasional day skipped (Knight 1908)" as quoted from Hunt and Flaspohler (1998).
egg on which female typically begins incubation–penultimate (1) or last (0)	penult	0	No specific information found. Assume incubation starts with last egg laid.
duration from start of incubation to hatch (days)	I	12	"Lasts 12–13 d from laying of last egg (Knight 1905, Harrison 1975)" as quoted from Hunt and Flaspohler (1998).

Parameter	Code	Value	Notes
duration from hatch to fledging of nestlings (days)	N	12	"Young depart nest 10–14 d after hatching (Knight 1905)" as quoted from Hunt and Flaspohler (1998). 10-12 d (Ehrlich et al. 1988). Assume 12 d is typical nestling phase duration.
duration since nest failure due to other reasons until female initiates new nest (days)	We	6	No specific data found on duration of We. "Nests lost to predation are sometimes replaced (Tufts 1986)" as quoted from Hunt and Flaspohler (1998). Assume that nests are replaced rapidly after failure and that a typical duration for We is 6 d.
duration since successful fledging until female initiates new nest (days)	Wf	100	No specific data found on duration of Wf. "In w. U.S., Dawson (1923) reported 2 broods to be rare, but see Dawson 1909 . In e. U.S., Nice (1926) reported a second nest after seeing fledglings from the first. Direct observation needed to confirm frequency of second broods" as quoted from Hunt and Flaspohler (1998). If renesting after success is rare, the duration of Wf needs to be long enough so that renesting not possible within egg laying period, so use default of 100 d.
female body weight (g) during breeding season	BdyWt	12	"Little variation between sexes and among subspecies. Mean mass (range) of birds on breeding grounds as follows (in g): *coronata* male, 12.9 (10.6–16.7; $n = 231$); female, 12.2 (9.9–15.3; $n = 29$); *auduboni* male, 12.3 (10.0–16.0; $n = 109$); female, 11.9 (10.0–14.0; $n = 79$; Dunning 1993)" as quoted from Hunt and Flaspohler (1998). Assume typical female weight is 12 g.
diet composition during breeding season		O	"During breeding season, mostly insects (adults and larvae) and other small invertebrates. Breeding-season diets included the following proportions of arthropods: East 78%, West 85% (15% "vegetable matter"; Forbush 1929). For first few days after chicks hatch, young are fed soft insects; then diet appears identical to that of adults (Knight 1908)" as quoted from Hunt and Flaspohler (1998). Also, adults consume shrub fruits and berries, especially fall & winter. Yellow-rumps are the only warbler that can consume waxy-coated fruits. Assume adult diet of 85% invertebrates and 15% fruit, while juveniles consume 100% invertebrates.
mean number of nest attempts/ female/ season			
mean number of successful broods/ female/ season			
mean number of fledglings/ successful nest	fpsn	1.11	No specific data found on the number of fledglings/successful nest. "At Mogollon Rim, AZ, 51.5%; percentage survival of each stage: egg (69.2%), incubation (81.4%), nestling (81.4%), average number of young fledged per nest 0.57 ± 1.28 SD (range 0–4. $n = 54$; T. Martin unpubl.)" as quoted from Hunt and Flaspohler (1998). If 0.57 fledglings per total number of nests and 51.5% of nests successful, then the mean of fledglings per successful nest would be 1.11.
mean number of fledglings/ female/ season (ARS)	ARS		

Reference list

Baily, A. M. and R. J. Niedrach. 1965. Birds of Colorado. Vol. 2. Denver Mus. Nat. Hist. Denver, CO.

Brandt, H. 1951. Arizona and its bird life: a naturalist's adventures with the nesting birds on the deserts, grasslands, foothills, and mountains of southeastern Arizona. The Bird Research Foundation, Cleveland, OH.

Dawson, W. L. 1909. The birds of Washington. Occidental Publ. Co., Seattle.

Dawson, W. L. 1923. The birds of California. Vol. 1. South Moulton Co., San Diego.

Dunning, Jr., J. B. 1993. CRC handbook of avian body masses. CRC Press, Boca Raton, FL.

Forbush, E. H. 1929. Birds of Massachusetts and other New England states. Pt. 3. Massachusetts Dept. Agriculture, Boston.

Gabrielson, I. N. and F. C. Lincoln. 1957. The birds of Alaska. Stackpole Co. Harrisburg, PA.

Harrison, H. 1975. A field guide to bird's nests. Houghton Mifflin, New York.

Hunt, P. D. and David J. Flaspohler. 1998. Yellow-rumped Warbler (*Setophaga coronata*). The Birds of North America Online (A. Poole, Ed.). Ithaca: Cornell Lab of Ornithology; Retrieved from the Birds of North America Online: http://bna.birds.cornell.edu.bnaproxy.birds.cornell.edu/bna/species/376.

Jewett, S. G., W. P. Taylor, W. T. Shaw, and J. W. Aldrich. 1953. Birds of Washington state. Univ. of Washington Press, Seattle.

Knight, O. W. 1905. Notes on the warblers found in Maine. *J. Maine Ornithol. Soc.* 7:71-76.

Knight, O. W. 1908. The birds of Maine. Charles Glass and Co., Bangor.

Nice, M. M. 1926. Behavior of blackburnian, myrtle, and black-throated blue warblers, with young. *Wilson. Bull.* 38:82-83.

Peck, G. K. and R. D. James. 1987. Breeding birds of Ontario: nidology and distribution, Vol. 2: Passerines. Royal Ontario Mus., Toronto, ON.

Tufts, R. W. 1986. Birds of Nova Scotia. Nimbus Publ. Ltd. and The Nova Scotia Museum, Halifax.

Wood, N. A. 1951. The birds of Michigan. Univ. of Michigan Press, Ann Arbor.

Cassin's sparrow (*Peucaea cassinii*)	Four-letter Alpha Code: CASP		
Species life-history parameters	Model Code	Typical value	Rationale
daily mortality rate during laying & incubation	m1	0.0305	"In Southern High Plains, Texas, Berthelsen and Smith (1995) found 46% of 30 nests in blue grama (*Bouteloua gracilis*)/sideoats grama fields successfully raised ≥1 young to fledging (1988–1989). At Tucson, AZ, of 19 nesting attempts in 1983, 10 were successful in rearing ≥1 young to fledging, 6 documented as failures, fate of 3 nests unknown (RKB)" as quoted from Dunning (1999). The weighted mean of these two studies is a 49% nest success rate. Over a 23 d nest period (i.e., 9+11+4-1), the daily nest mortality rate is 0.0305.
daily mortality rate during nestling-rearing	m2	0.0305	
date of first egg of first nest (dd-mmm)	T1	18-May	"Range of egg dates throughout range (from north to south; Hubbard 1977 unless otherwise noted): Nebraska: 30 Jun (Bock and Scharf 1994); Kansas: May–Jun; Colorado: 2 nests 16 May and 14 Jun, young observed 30 Jun (Bailey and Niedrach 1965, Kingery and Julian 1971); Oklahoma: 26 May–22 Jul; Texas: 1 Mar–1 Aug, with 52% of 85 clutches falling in May; New Mexico: 3 nests in late Jun and early Jul with fledglings on 2 Jul; Arizona: late Jul–early Sep (Monson and Phillips 1981)" as quoted from Dunning (1999).
date of first egg of last nest (dd-mmm)	Tlast	20-Jul	
length of rapid follicle growth period (RFG) for each egg (days)	rfg	3	Based on allometric equation (RFG = $2.852*$Egg mass$^{0.31}$) from Alisauskas and Ankney (1992), using an egg mass of 1.6 g, assume rfg = 3 d.
mean clutch size	clutch	4	"In se. Arizona, mean clutch size 3.3 eggs ± 0.48 SD (range 3–4, $n = 22$; RKB). On fields enrolled in Conservation Reserve Program, Southern High Plains, Texas, 4.4 eggs ± 0.61 SE ($n = 34$; Berthelsen and Smith 1995). In Oklahoma, 4.2 eggs (range 3–5, $n = 6$; Sutton 1967). In Nebraska, 3 nests with 5 eggs each (Bock and Scharf 1994)" as quoted from Dunning (1999).
mean intra-egg laying interval (days)	eli	1	"Schnase et al. (1991): female lays 1 egg in the morning, beginning 2–3 d after nest construction" as quoted from Dunning (1999).
egg on which female typically begins incubation—penultimate (1) or last (0)	penult	1	"Incubation begins with penultimate egg" as quoted from Dunning (1999).
duration from start of incubation to hatch (days)	I	11	"Three nests in se. Arizona: ≥11 d, ≥11 d, ≥9 d (RKB). One nest in Texas, 11 d (Schnase et al. 1991)" as quoted from Dunning (1999).
duration from hatch to fledging of nestlings (days)	N	9	"Young in 4 nests followed in se. Arizona fledged 7, 7, 8, and 9 d after hatching (RKB); several of these nests may have fledged prematurely due to observer presence. Schnase et al. (1991) followed 1 nest that fledged naturally at 9 d" as quoted from Dunning (1999).
duration since nest failure due to other reasons until female initiates new nest (days)	We	6	No specific information on the duration of We. Dunning et al. (1999) report that renesting after nest loss occurs. Assume females renest rapidly after nest loss, with a typical value for We of 6 d.

duration since successful fledging until female initiates new nest (days)	Wf	100	No specific information on the duration of Wf. "Limited studies document only 1 brood/season, but skylarking by male after fledging of first brood suggest double-brooding is at least possible (Schnase et al. 1991)" as quoted from Dunning et al. (1999). If we assume that females typically quit breeding after a successful brood, the value for Wf would be long enough to prevent a new nest attempt, so choose a default of 100 d.
female body weight (g) during breeding season	BdyWt	18.1	"Breeding birds, means, Jul–Sep, se. Arizona, (RKB): male, 17.8 g ± 1.19 SD (range 16.0–19.5, n = 28); female, 18.1 g ± 1.25 SD (range 16.0–21.5, n = 15)" as quoted from Dunning et al. (1999).
diet composition during breeding season		I	"Insects during the nesting season, weed and grass seeds during nonbreeding season. Stomach contents of 10 adults collected in late Jun–early Jul included 52% animal, 48% plant material (Wolf 1977). Oberholser (1974) and Wolf (1977) listed the following diet items: grasshoppers (Orthoptera); caterpillars (Lepidoptera); true bugs (Hemiptera); ants, bees and wasps (Hymenoptera); weevils (Coleoptera); spiders (Arachnida); and snails (Gastropoda) in warm-weather months. At 1 nest in se. Arizona, 197 of 208 prey items delivered to young by parents were grasshoppers (K. Jepsen-Innes unpubl.)" as quoted from Dunning et al. (1999). Although some seeds may be consumed, assume during breeding season both adults and juveniles consume 100% invertebrates.
mean number of nest attempts/ female/ season			
mean number of successful broods/ female/ season			
mean number of fledglings/ successful nest	fpsn	2.2	"A total of 13 fledglings were produced by the six males in this study" as quoted from Schnase et al. (1991). Since only one successful brood per pair was produced during the breeding season, if we interpret this passage to mean that each of the 6 males produced a successful brood, then 2.2 fledglings were produced per successful nest.
mean number of fledglings/ female/ season (ARS)	ARS		

Reference list

Bailey, A. H. and R. J. Niedrach. 1965. Birds of Colorado. Vol. 2. Denver Mus. Nat. Hist. Denver.

Berthelsen, P. S. and L. M. Smith. 1995. Nongame bird nesting on CRP lands in the Texas Southern High Plains. *J. Soil Water Conserv.* 50:672-675.

Bock, C. E. and W. C. Scharf. 1994. A nesting population of Cassin's Sparrows in the sandhills of Nebraska. *J. Field Ornithol.* 65:472-475.

Dunning Jr., J. B., R. K. Bowers Jr., S. J. Suter, and C. E. Bock. 1999. Cassin's Sparrow (*Peucaea cassinii*), The Birds of North America Online (A. Poole, Ed.). Ithaca: Cornell Lab of Ornithology; Retrieved from the Birds of North America Online: http://bna.birds.cornell.edu.bnaproxy.birds.cornell.edu/bna/species/471.

Hubbard, J. P. 1977. The status of Cassin's Sparrow in New Mexico and adjacent states. *Am. Birds* 31:933-941.

Kingery, H. E. and P. R. Julian. 1971. Cassin's Sparrow parasitized by cowbird. *Wilson Bull.* 83:439.

Monson, G. and A. Phillips. 1981. Annotated checklist of the birds of Arizona. Univ. of Arizona Press, Tucson.

Oberholser, H. C. 1974. The bird life of Texas. Univ. of Texas Press, Austin.

Schnase, J. L., W. E. Grant, T. C. Maxwell, and J. J. Leggett. 1991. Time and energy budgets of Cassin's Sparrow (*Aimophila cassinii*) during the breeding season: evaluation through modelling. *Ecol. Model.* 55:285-319.

Sutton, G. M. 1967. Oklahoma birds: their ecology and distribution, with comments on the avifauna of the southern Great Plains. Univ. of Oklahoma Press, Norman.

Wolf, L. L. 1977. Species relationships in the avian genus *Aimophila*. *Ornithol. Monogr.* 23

Chipping sparrow (*Spizella passerina*)

Four-letter Alpha Code: CHSP

Species life-history parameters	Model Code	Typical value	Rationale
daily mortality rate during laying & incubation	m1	0.027	"Reproductive success varies from year to year. At Guelph, ON, 48.4% (range 19.1–66.7) of nests produced at least 1 fledgling ($n = 8$ seasons, 381 nests in which incubation began). Success rates (i.e., nests producing at least 1 Chipping Sparrow fledgling) in Minnesota vary from 17.4 to 50% (Keller 1979, Buech 1982, Albrecht and Oring 1995); 62% at Battle Creek, MI (Walkinshaw 1944); and 63.6% in Algonquin Park, ON (Reynolds and Knapton 1984)" as quoted in Middleton (1998). Assume a 50% nest success rate is typical – this translates to a daily nest mortality rate over 25 d (i.e., 11+11+4-1) of 0.027.
daily mortality rate during nestling-rearing	m2	0.027	
date of first egg of first nest (dd-mmm)	T1	Apr-30	"In general, breeding season is protracted; this species regularly produces 2 broods annually, but rarely 3 (Walkinshaw 1952, Keller 1979, Peck and James 1987, Scott and Lemon 1996, ALAM). First egg dates vary with latitude: in California, 24 Mar (peak 10–30 May; Johnson 1968); Ohio, 16 Apr (peak late Apr–early May; Peterjohn 1989); central Michigan, 8 May (peak early–mid-May; Walkinshaw 1944); nw. Minnesota, 19 May (peak late May–early Jun; Keller 1979); n. Ontario, 7 Jun (peak for Ontario, 4–20 Jun; Speirs 1985, Peck and James 1987); Mackenzie District, Northwest Territories, 9 Jun (Stull 1968). Earliest date for first egg at Guelph, ON, 7 May (mean 12.8 May ± 4.5 d SD; $n = 15$)" as quoted in Middleton (1998). Assume typical value for T1 of April 30.
date of first egg of last nest (dd-mmm)	Tlast	Jul-10	"Last egg date more difficult to determine; varies with geographic location from latter half of Jul to late Aug (Stull 1968, Keller 1979, Baumgartner and Baumgartner 1992). Latest known egg date at Guelph, ON, 15 Jul ($n = 8$ seasons); latest record for Ontario, 14 Aug (Peck and James 1987)" as quoted in Middleton (1998). Based on Figure 4 in Middleton (1998), adjusting for start of laying. Assume typical value for Tlast of July 10.
length of rapid follicle growth period (RFG) for each egg (days)	rfg	3	Based on allometric equation from Alisauskas and Ankney (1992), assume rfg = 3 d.
mean clutch size	clutch	4	"At Guelph, ON, mean clutch size is 3.7 eggs ± 0.65 SD (range 2–5, $n = 377$ unparasitized clutches with complete data; ALAM), similar to clutch size in central Michigan and ne. Pennsylvania (Walkinshaw 1944, Stull 1968; Table 1). Many birds attempt second broods; clutches in repeat and second nests appear to be smaller, but not confirmed (ALAM). Seasonal decline in clutch size from 3.8 eggs in May to 3.0 in Jul at Battle Creek, MI ($n = 45$; Walkinshaw 1944); similar decline at Guelph, ON (ALAM)" as quoted in Middleton (1998).
mean intra-egg laying interval (days)	eli	1	"Normal laying sequence is 1 egg/d until clutch is complete (Bradley 1940, Walkinshaw 1944, Stull 1968, Reynolds and Knapton 1984, ALAM)" as quoted in Middleton (1998).
egg on which female typically begins incubation–penultimate (1) or last (0)	penult	1	"Only female incubates, beginning with laying of penultimate egg (Harrison 1978, Keller 1979, Reynolds and Knapton 1984, ALAM)" as quoted in Middleton (1998).

Description	Code	Value	Notes
duration from start of incubation to hatch (days)	I	11	"Usually given as 10–12 d (range 10–15), but may be as short as 7 d (Bradley 1940, Walkinshaw 1952, Dawson and Evans 1957, Harrison 1957, Harrison 1978, Keller 1979, Reynolds and Knapton 1984, Peck and James 1987)" as quoted in Middleton (1998).
duration from hatch to fledging of nestlings (days)	N	11	"Nest departure possible any time from day 8 on (Weaver 1937, Dawson and Evans 1957), but at Guelph, ON, and elsewhere, most young depart nest at age 9–12 d (Bradley 1940, Walkinshaw 1944, Harrison 1978, Keller 1979, Reynolds and Knapton 1984, ALAM). Survival of young departing nest before day 9 at Guelph, ON, is greatly reduced; if undisturbed, young stay in nest as long as possible (ALAM)" as quoted in Middleton (1998).
duration since nest failure due to other reasons until female initiates new nest (days)	We	5	No specific data on duration of We. Assume that renesting after failure occurs rapidly and that a typical duration for We is 5 d.
duration since successful fledging until female initiates new nest (days)	Wf	9	"Young are dependent on parents for about 3 wk, but precise timing of independence difficult to determine (Walkinshaw 1944, Peterjohn 1989, ALAM). Where female attempts second brood, male assumes major responsibility for chicks from first nest until they reach independence (Keller 1979, Middleton and Prescott 1989, ALAM). Mean of 9.3 d ± 5.99 SD (range 4–17, $n = 6$) between nest departure at first nest and first egg of second clutch (Keller 1979)" as quoted in Middleton (1998).
female body weight (g) during breeding season	BdyWt	13.0	"At Guelph, ON, females are heavier than males during May and Jun, probably because of egg-laying (ALAM). Mean mass of males at Guelph, ON: in May, 12.1 g ± 0.55 SD (range 11.5–13.5, $n = 29$); Jun, 12.0 g ± 0.69 SD (range 11.0–13.0, $n = 11$). Of females: in May, 13.4 g ± 1.09 SD (range 11.5–15.5, $n = 23$); Jun, 12.3 g ± 0.53 SD (range 11.5–13.0, $n = 8$). For large sample in Pennsylvania, but without sex and date information, mean 12.3 g ± 0.84 SD (range 9.8–18.8, $n = 934$; Dunning 1993)" as quoted in Middleton (1998).
diet composition during breeding season	O		"Observation (Judd 1900, 1901, Forbush 1913, Pulliam 1980) and stomach content analysis ($n = 250$ [Judd 1901]; $n = 29$ [Allaire and Fisher 1975], $n = 46$ [Pulliam 1980]) show that Chipping Sparrow consumes seeds throughout year and that invertebrates, primarily insects, form major part of diet during breeding season. Specific foods consumed may vary with locality, but basic diet appears similar across North America (Judd 1901). Summer and early-fall foods consist of 62% plant matter, primarily grass seeds, and 38% invertebrates, primarily insects (Judd 1901). Details of nestling diet are lacking, but in early stages appears to include mostly seeds, and quantity of invertebrate food increases as young mature" as quoted in Middleton (1998). Assume diet of 62% seeds and 38% invertebrates for adults and 80% seeds and 20% invertebrates for juveniles.
mean number of nest attempts/ female/ season			

mean number of successful broods/ female/ season			"Many studies report double broods (Walkinshaw 1952, Sutton 1960, Keller 1979, Peterjohn 1989, Scott and Lemon 1996, ALAM); others report none (Reynolds and Knapton 1984, Albrecht and Oring 1995). Frequency of double broods reported at 12% in Minnesota (Keller 1979), 24% in Ontario (Scott and Lemon 1996), and from "several" to "most pairs" in Michigan (Walkinshaw 1952, Sutton 1960). Likelihood of double brood depends on success of an early nest and on ability of male to care for first brood (Keller 1979, ALAM)" as quoted in Middleton (1998).
mean number of fledglings/ successful nest	fpsn	2.1	In ON, 279 fledglings were produced from 159 successful nests (Middleton 1998). In MI, 93 fledglings produced from 31 successful nests (Walkinshaw 1944). In ON, 50 fledglings produced from 14 successful nests (Reynolds and Knapton 1984). Weighted mean of 3 studies (i.e., 422/204) is 2.1 fledglings/successful nest.
mean number of fledglings/ female/ season (ARS)	ARS	3.0	"In Minnesota, double-brooded pairs produced mean of 4.5 young/season ($n = 4$) compared to 2.6 for single-brooded pairs ($n = 15$; Keller 1979)" as quoted in Middleton (1998). Weighed mean of 3.0 fledglings/female in population.

Note: "*Condition at Departure.* Well feathered; body mass between 80 and 90% of adult mass; body growth about 90% complete, with greatest growth still to occur in bill length (Weaver 1937, Walkinshaw 1944, Dawson and Evans 1957, Reynolds and Knapton 1984" as quoted in Middleton (1998).

Reference List

Albrecht, D. J. and L. W. Oring. 1995. Song in Chipping Sparrows, *Spizella passerina*, structure and function. *Anim. Behav.* 50:1233-1241.

Alisauskas, R. T., and C. D. Ankney. 1992. The cost of egg laying and its relationship to nutrient reserves in waterfowl. In: *Ecology and Management of Breeding Waterfowl.* B. Batt (Ed.), University of Minnesota Press. Pp 30-61.

Allaire, P. N. and C. D. Fisher. 1975. Feeding ecology of three resident sympatric sparrows in eastern Texas. *Auk* 92:260-269.

Baumgartner, F. M. and A. M. Baumgartner. 1992. Oklahoma bird life. Univ. of Oklahoma Press, Norman.

Bradley, H. L. 1940. A few observations on the nesting of the Eastern Chipping Sparrow. *Jack-Pine Warbler* 18:35-46.

Buech, R. R. 1982. Nesting ecology and cowbird parasitism of Clay-Colored, Chipping, and Field Sparrows in a Christmas tree plantation. *J. Field Ornithol.* 53:363-369.

Dawson, W. R. and F. C. Evans. 1957. Relation of growth and development to temperature regulation in nestling Field and Chipping Sparrows. *Physiol. Zool.* 30:315-327.

Dunning, Jr., J. B. 1993. CRC handbook of avian body masses. CRC Press, Boca Raton, FL.

Forbush, E. H. 1913. Useful birds and their protection. 4th ed. Mass. State Board of Agric., Boston.

Harrison, C. 1978. A field guide to the nests, eggs and nestlings of North American birds. Collins Sons and Co., Ltd. Glasgow, UK.

Keller, M. E. 1979. Breeding behavior and reproductive success of Chipping Sparrows in northwestern Minnesota. Master's Thesis. Univ. of North Dakota, Grand Forks.

Johnson, R. R. 1968. _Spizella passerina arizonae_ Coues. Western Chipping Sparrow. Pages 1184-1186 _in_ Life histories of North American cardinals, grosbeaks, buntings, towhees, finches, sparrows, and allies. (O. L. Austin Jr., Ed.) _U.S. Natl. Mus. Bull._ no. 237 (2)

Judd, S. D. 1900. The food of nestling birds. Pages 411-436 _in_ U.S. Dep. Agric. Yearbook.

Judd, S. D. 1901. The relation of sparrows to agriculture. _U.S. Dep. Agric. Biol. Surv._ 15:76-78.

Middleton, A. L. 1998. Chipping Sparrow (Spizella passerina), The Birds of North America Online (A. Poole, Ed.). Ithaca: Cornell Lab of Ornithology; Retrieved from the Birds of North America Online: http://bna.birds.cornell.edu.bnaproxy.birds.cornell.edu/bna/species/334.

Middleton, A. L. A. and D. R. C. Prescott. 1989. Polygyny, extra-pair copulations, and nest helpers in the Chipping Sparrow. _Spizella passerina. Can. Field-Nat._ 103:61-64.

Peck, G. K. and R. D. James. 1987. Breeding birds of Ontario: nidiology and distribution, Vol. 2: passerines. R. Ont. Mus. Life Sci. Misc. Publ., Toronto.

Peterjohn, B. G. 1989. The birds of Ohio. Indiana Univ. Press, Bloomington.

Pulliam, H. R. 1980. Do Chipping Sparrows forage optimally? _Ardea_ 68:75-82.

Reynolds, J. D. and R. W. Knapton. 1984. Nest-site selection and breeding biology of the Chipping Sparrow. _Wilson Bull._ 96:488-493.

Scott, D. M. and R. E. Lemon. 1996. Differential reproductive success of Brown-headed Cowbirds with Northern Cardinals and three other hosts. _Condor_ 98:259-271.

Speirs, J. M. 1985. Birds of Ontario. Vol. 2. Nat. Heritage/Nat. Hist. Inc. Toronto.

Stull, W. D. 1968. _Spizella passerina_ (Bechstein): Eastern and Canadian Chipping Sparrows. Pages 1166-1184 _in_ Life histories of North American cardinals, grosbeaks, buntings, towhees, finches, sparrows, and allies. (O. L. Austin Jr., Ed.) _U.S. Natl. Mus. Bull._ no. 237(2)

Sutton, S. M. 1960. The nesting fringillids of the Edwin S. George Reserve, southeastern Michigan (Part VI). _Jack-Pine Warbler_ 38:46-65.

Walkinshaw, L. H. 1944. The Eastern Chipping Sparrow in Michigan. *Wilson Bull.* 56:193-205.

Walkinshaw, L. H. 1952. Chipping Sparrow notes. *Bird-Banding* 23:101-108.

Weaver, R. 1937. Measurement of growth in the Eastern Chipping Sparrow. *The Auk* 54:103-104.

Field sparrow (*Spizella pusilla*)			Four-letter Alpha Code: FISP
Species life-history parameters	Model Code	Typical value	Rationale
daily mortality rate during laying & incubation	m1	0.037	In PA from 1987-2006, 45% ± 12.8 (n=939) of nest fledged at least one young (Carey et al. 2008). In MI from 1938-1950, 38% ± 10.1 (n=613) of nests were successful. Comparable fledging success rates have been found in other short term studies (44% - Batts 1961; 35% - Crooks 1948; 27% - Nolan 1963). In an Illinois population, however, only 10% of nests successfully fledged young (Best 1978). Based on apparent nest success in large studies in PA and MI , the weighted mean = 42%, which translates to a daily nest mortality rate of 0.037 over a 23 d nest period (i.e., 4+11+8).
daily mortality rate during nestling-rearing	m2	0.037	
date of first egg of first nest (dd-mmm)	T1	May-15	"In Missouri, first eggs (including renestings) laid from 29 Apr–10 Aug; earliest chick hatch on 15 May; latest fledging 7 Sep (DEB). In Michigan (1939-1948) earliest first egg dates ranged from 29 Apr – 16 May (Walkinshaw 1978). In Pennsylvania (1987-2006), earliest first egg dates (n = 905) ranged from 3 May – 16 May; latest first egg dates ranged from 4 Jul – 27 Jul" as quoted in Carey et al. (1994). Based on Figure 4 in Carey et al. (2008) assume core of egg laying period is May 15 to July 1.
date of first egg of last nest (dd-mmm)	Tlast	Jul-1	
length of rapid follicle growth period (RFG) for each egg (days)	rfg	3	Based on allometric equation (RFG = 2.852*Egg mass ^0.31) from Alisauskas and Ankney (1992), using and egg mass of 1.69 g, assume rfg = 3 d.
mean clutch size	clutch	4	In PA, mean of 3.69 ± 0.57 SD, mode = 4, n=158 and in MO, mean of 3.96 ± 0.59 SD, mode=4, n=47 (Carey et al. 2008).
mean intra-egg laying interval (days)	eli	1	"One egg/d (Walkinshaw 1968a, MC)" as quoted in Carey et al. (2008).
egg on which female typically begins incubation–penultimate (1) or last (0)	penult	0	"In two-egg clutches female begins incubating with the last egg; 3-4 eggs - incubation begins the night before the last egg is laid; 5 eggs – 2 nights before last egg (Walkinshaw 1978). During cold early spring conditions, some females delay start of incubation up to 4 d after laying the last egg (MC)" as quoted in Carey et al. (2008).
duration from start of incubation to hatch (days)	I	11	"11–12 d; occasionally 10; as long as 17 d in cool wet springs when onset of incubation behavior may be delayed (Walkinshaw 1936, 1968a, 1978; Best 1978, MC)" as quoted in Carey et al. (2008).
duration from hatch to fledging of nestlings (days)	N	8	"Typically 7–8 d; as early as 5 d if disturbed (Walkinshaw 1968a, MC)" as quoted in Carey et al. (2008).
duration since nest failure due to other reasons until female initiates new nest (days)	We	5	"Egg laying in a new nest starts about 5 d after loss or desertion. (Walkinshaw 1939, MC)" as quoted in Carey et al. (2008).
duration since successful fledging until female initiates new nest (days)	Wf	12	"Females fledging young begin laying in a new nest 6–20 d after fledging (Walkinshaw 1968a, DEB, MC)" as quoted in Carey et al. (2008). Assume 12 d as typical duration for Wf.

female body weight (g) during breeding season	BdyWt	13.0	"No significant difference between sexes in Pennsylvania during the breeding season (MC). Males: n = 15, range 11.5–14.3 g, mean = 13.1 ± 0.67 SD g. In s. New York, breeding males: n = 23, range 12.0–15.0 g, mean = 13.2 ± 0.85 SD g (DAN); females: n = 17, range 11.4–14.0 g, mean = 13.0 ± 0.72 SD g" as quoted in Carey et al. (2008).
diet composition during breeding season		O	"Winter: > 90% of food items found in analyses of gut contents were seeds, virtually all from various grasses. With approach of spring and through the summer, the proportion of plant food in diet declines. Grass seeds are <50% of the summer diet; insects the rest; seed proportion rises once again from August through fall (Judd 1901, Martin et al. 1951, Martin et al. 1951, Evans 1964, Pulliam and Enders 1971, Allaire and Fisher 1975)" as quoted in Carey et al. (2008). Similarly, Martin et al. show spring and summer diet approximately 50% seeds and 50% invertebrates. Nestlings fed almost exclusively invertebrates.
mean number of nest attempts/ female/ season			"In Iowa, for pairs mated all season, there was an average of 4 nesting attempts/pair (Crooks and Hendrickson 1953); in Pennsylvania a mean of 2.9 ± 0.91 SD nesting attempts/season/female (range 1–5, n = 98; MC)" as quoted in Carey et al. (2008).
mean number of successful broods/ female/ season		1.09	In PA from 1987–1993, 46% of females fledged 1 brood, 30% fledged 2 broods, 1% fledged 3 broods, and 23% failed to fledge a brood (Carey et al. 1994). Average of 1.09 successful broods/female.
mean number of fledglings/ successful nest	fpsn	3.07	Nests typically fledge 3 or 4 young. In PA, mean of 3.0 ± 0.89 SD, range = 1–5, n=172 and in MO, mean of 3.4 ± 0.82 SD, range = 1–5, n=39 (Carey et al. 1994). Weighted mean of two studies = 3.07 fledglings/successful nest.
mean number of fledglings/ female/ season (ARS)	ARS	3.0	In Pennsylvania from 1987–2006, 312 of 430 breeding females (73%) fledged at least one young each year; the average female fledged 3.0 (± 0.8 SD; n=430) young per year (Carey et al. 2008). Ricklefs and Bloom (1977) estimated 6.17 fledglings/female/yr, but used a longer egg laying period.

Note: Young weigh about 10.5 g when ready to leave nest (Walkinshaw 1936). Four 13-d-old fledglings in Michigan: mean mass = 11.7 g (Walkinshaw 1978). Young birds are comparable in size to adults by 5-6 weeks old (Walkinshaw 1978).

Reference List

Alisauskas, R. T., and C. D. Ankney. 1992. The cost of egg laying and its relationship to nutrient reserves in waterfowl. In: *Ecology and Management of Breeding Waterfowl.* B. Batt (Ed.), University of Minnesota Press. Pp 30-61.

Allaire, P. N. and C. D. Fisher. 1975. Feeding ecology of three resident sympatric sparrows in eastern Texas. *Auk* 92:260-269.

Batts, Jr., H. L. 1961. Nesting success of birds on a farm in southern Michigan. *Jack-Pine Warbler* 39(2):72-83.

Best, L. B. 1978. Field Sparrow reproductive success and nesting ecology. *Auk* 95:9-22.

Carey, M., M. Carey, D. E. Burhans and D. A. Nelson. 2008. Field Sparrow (*Spizella pusilla*), The Birds of North America Online (A. Poole, Ed.). Ithaca: Cornell Lab of Ornithology; Retrieved from the Birds of North America Online: http://bna.birds.cornell.edu.bnaproxy.birds.cornell.edu/bna/species/103

Crooks, M. P. 1948. Life history of the Field Sparrow *Spizella pusilla pusilla* (Wilson). Master's Thesis. Iowa State College, Cedar Falls.

Crooks, M. P. and G. O. Hendrickson. 1953. Field Sparrow life history in central Iowa. *Iowa Bird Life* 23:10-13.

Evans, F. C. 1964. The food of Vesper, Field, and Chipping Sparrows nesting in an abandoned field in southeastern Michigan. *Amer. Midl. Nat.* 72:57-75.

Judd, S. D. 1901. The relation of sparrows to agriculture. *U.S. Dept. Agric., Div. Biol. Survey, Bull.* No. 15.

Martin, A. C., H. S. Zim, and A. L. Nelson. 1951. American wildlife and plants. McGraw-Hill, New York.

Nolan, Jr., V. 1963. Reproductive success of birds in a deciduous scrub habitat. *Ecology* 44:305-313.

Pulliam, H. R. and F. Enders. 1971. The feeding ecology of five sympatric finch species. *Ecology* 52:557-566.

Ricklefs, R. E., and G. Bloom. 1977. Components of avian breeding productivity. *The Auk* 94: 86-96.

Walkinshaw, L. H. 1936. Notes on the Field Sparrow in Michigan. *Wilson Bull.* 48:94-101.

Walkinshaw, L. H. 1939. Nesting of the Field Sparrow and survival of the young. *Bird-Banding* 10:107-114, 149-157.

Walkinshaw, L. H. 1968a. Eastern Field Sparrow. Pages 1217-1235 *in* Life histories of North American cardinals, grosbeaks, buntings, towhees, finches, sparrows, and allies. (O. L. Austin, Ed.). *U.S. Natl. Mus. Bull.* No. 237.

Walkinshaw, L. H. 1978. Life history of the eastern Field Sparrow in Calhoun County, Michigan. University Microfilm International, Ann Arbor.

Vesper sparrow (*Pooecetes gramineus*)

Four-letter Alpha Code: VESP

Species life-history parameters	Model Code	Typical value	Rationale
daily mortality rate during laying & incubation	m1	0.067	In IA, Patterson and Best (1996) reported an apparent nest success of 33% but a nest success rate calculated based on the Mayfield method of only 16% (lowest in Table 5 of Jones and Cornely 2002), with daily nest survival rates during egg and nestling stages of 0.918 and 0.943, respectively. In IA, Rodenhouse and Best (1983) reported an apparent nest success of 29%, but using Mayfield they reported overall nest success rate is 13%, resulting in a daily nest survival rate over a 27 d nesting period (i.e., 10+13+4) of 0.927. In WV, Wray et al. (1982) reported an apparent nest success over 3 yr of 31%, but using the Mayfield method the overall nest success rate was 18%, or a daily nest survival rate of 0.938. In WA, Vander Haegen (as cited in Jones and Cornely 2002) reported an overall nest success of 25.4% and daily nest survival rate of 0.95. The mean of these four studies gives daily nest survival rate for egg and nestling stages of 0.933 and 0.940, respectively, and corresponding daily nest mortality rates of 0.067 and 0.060.
daily mortality rate during nestling-rearing	m2	0.060	
date of first egg of first nest (dd-mmm)	T1	4-May	Table 3 of Jones and Cornely (2002) presents median, mean, peak, and range of clutch initiation dates from 8 studies. To estimate typical egg laying dates, the mean values were calculated for the start and end of the ranges, although the start date of the range from the OR study (i.e., May 31) was removed since it came after the reported median and peak and probably represents an error.
date of first egg of last nest (dd-mmm)	Tlast	19-Jul	
length of rapid follicle growth period (RFG) for each egg (days)	rfg	4	Based on allometric equation (RFG = 2.852*Egg mass ^0.31) from Alisauskas and Ankney (1992), using and egg mass of 2.7 g, assume rfg = 4 d.
mean clutch size	clutch	4	Average 3.6 (range 2-6) (cited in Jones and Cornely 2002)
mean intra-egg laying interval (days)	eli	1	Probably 1 egg laid/day (Jones and Cornely 2002).
egg on which female typically begins incubation–penultimate (1) or last (0)	penult	0	No information in Jones and Cornely (2002), so assume starts with last egg.
duration from start of incubation to hatch (days)	I	13	12 to 13 d (range 11-14 d) (Jones and Cornely 2002).
duration from hatch to fledging of nestlings (days)	N	10	Departure at 9.6 d of age (range 7-14; n=96; Dawson and Evans 1960) as cited in Jones and Cornely (2002).
duration since nest failure due to other reasons until female initiates new nest (days)	We	7	While a high percentage of lost nests are lost to predators and farm operations and renesting is common after failure, there is little information on the duration of We. Assume females renest relatively rapidly after failure with a We of 7 d.
duration since successful fledging until female initiates new nest (days)	Wf	7	Fledglings "dependent on adults for 20-29 d after fledging; male may feed and care for first brood while female renests (Perry and Perry 1918)" as cited in Jones and Cornely (2002), but Perry paper offers little evidence for this statement. Assume females leave fledgling care to male and renest relatively rapidly after success with a Wf of 7 d.

female body weight (g) during breeding season	BdyWt	23	Adult females of four subspecies average about 23 g (Jones and Cornely 2002).
diet composition during breeding season		O	Adults consume mostly insects and small seeds (Jones and Cornely 2002). In summer, 56% invertebrates and 44% weed seeds (Martin et al. 1951). "Seeds rare in nestling diet; insects regularly provided" (Jones and Cornely 2002). Assume juveniles consume 100% invertebrates.
mean number of nest attempts/ female/ season			In WV mean of 3.6 nesting attempts/female/yr (range 3.4 to 4.1, n=70) (Wray et al. 1982).
mean number of successful broods/ female/ season			
mean number of fledglings/ successful nest	fpsn	3.0	In WV, mean number of fledglings/successful nest 3.0 (range 3.0-3.1 over 3 yr, n=70) (Wray et al. 1982).
mean number of fledglings/ female/ season (ARS)	ARS	1.9	Although Wray et al. (1982) reported in Table 4 the mean number of fledglings per female over 3 yr study based on the formula in Pinkowski (1979), it appears that they used mean clutch size rather than the mean number of fledglings per successful nest. By using the data presented in Wray et al. (1982), the mean ARS was recalculated using Pinkowski's formula over three yrs of 3.4 fledglings/female when using the apparent nest success rate or 1.9 fledglings/female when using the Mayfield-based estimate of nest success.

Note: With vesper sparrows there is a large discrepancy between estimates of apparent nest success rates and those calculated using the Mayfield method, possibly indicating a bias toward finding nests late in the nesting period.

Reference list

Alisauskas, R. T., and C. D. Ankney. 1992. The cost of egg laying and its relationship to nutrient reserves in waterfowl. In: Ecology and Management of Breeding Waterfowl. B. Batt (Ed.), University of Minnesota Press. Pp 30-61.

Dawson, W. R. and F. C. Evans. 1960. Relation of growth and development to temperature regulation in nestling Vesper Sparrows. Condor 62:329-340.

Jones, S. L. and J. E. Cornely. 2002. Vesper Sparrow (Pooecetes gramineus), The Birds of North America Online (A. Poole, Ed.). Ithaca: Cornell Lab of Ornithology; Retrieved from the Birds of North America Online: http://bna.birds.cornell.edu.bnaproxy.birds.cornell.edu/bna/species/624.

Martin, A. C., H. S. Zim, and A. L. Nelson. 1951. American wildlife and plants: A guide to wildlife food habits. Dover Publications, Inc., New York. 500 pp.

Patterson, M. P. and L. B. Best. 1996. Bird abundance and nesting success in Iowa CRP fields: the importance of vegetation structure and composition. Am. Midl. Nat. 135:153-176.

Perry, E. M. and W. A. Perry. 1918. Home life of the Vesper Sparrow and the Hermit Thrush. *Auk* 35:310-321.

Pinkowski, B. C. 1979. Annual productivity and its measurement in a multi-brooded passerine, the eastern bluebird. *Auk* 96:562-572.

Rodenhouse, N. L. and L. B. Best. 1983. Breeding ecology of Vesper Sparrow in corn and soybean fields. *Am. Midl. Nat.* 110:265-275.

Wray II, T., K. A. Strait, and R. C. Whitmore. 1982. Reproductive success of grassland sparrows on a reclaimed surface mine in West Virginia. *Auk* 99:157-164.

Lark sparrow (*Chondestes grammacus*)

Four-letter Alpha Code: LASP

Species life-history parameters	Model Code	Typical value	Rationale
daily mortality rate during laying & incubation	m1	0.031	"In n. Illinois grasslands, for 30 nests monitored between 1995 and 1997, daily survival probability was 0.9310 with Mayfield nest success about 22.3% with a nest cycle of 21 d (J. R Herkert unpubl.)" as quoted from Martin & Parrish (2000). This is equivalent to a daily nest mortality rate of 0.069. In OK, apparent nest success of 42% (14 of 33 nests fledged ≥1 nestling; data from Table 1 in Newman 1970), which is equivalent to a daily nest mortality rate of 0.031 over 27 d (i.e., 12+12+4-1) nest period.
daily mortality rate during nestling-rearing	m2	0.031	
date of first egg of first nest (dd-mmm)	T1	1-May	"Dates of clutch initiation from all locations ranged from 23 Mar–26 Jul (McNair 1985). Mean clutch initiation date for all geographic areas based on museum oology collections is 14 May ± 21 d SD (*n* = 919); based on Nest Record Cards (NRC) 1 Jun ± 22 d (*n* = 209). Earliest mean dates in Texas: 9 May (range 29 Apr–20 May, *n* = 210) based on museum data; 24 May (range 7 May–7 Jun, *n* = 71) based on NRC data. Latest dates in n. U.S. and Canada: 13 Jun (range 2–19 Jun, *n* = 5) based on museum data; 9 Jun (range 4–19 Jun, *n* = 23) based on NRC data, indicating a latitudinal gradient in egg dates (McNair 1985). Northern latitudes shown a narrower window for egg-laying dates than middle or southern latitudes; Oklahoma, 14 Apr–14 Jul; Kansas, 1 May–20 Jul; N. Dakota, 20 Jun–11 Jul (Johnsgard 1979)" as quoted from Martin and Parrish (2000). Based on Figure 4 in Martin and Parrish (2000), assume typical egglaying period is from May 1 to June 30.
date of first egg of last nest (dd-mmm)	Tlast	30-Jun	
length of rapid follicle growth period (RFG) for each egg (days)	rfg	4	RFG period for passerines typically 3-4 d.
mean clutch size	clutch	4	"Mean for all geographic areas 4.09 ± 0.66 SD (*n* = 928) based on museum oology collections and 3.84 ± 0.70 SD (*n* = 209) based on NRC data; range from 3–6 eggs; mode 4, with clutches of 5 more frequent than 3 (McNair 1985)" as quoted from Martin and Parrish (2000).
mean intra-egg laying interval (days)	eli	1	"Egg-laying at Marshall Co., OK, occurred between 05:00 and 07:00 (Baepler 1968)" as quoted from Martin and Parrish (2000). Presumably 1 d/egg.
egg on which female typically begins incubation–penultimate (1) or last (0)	penult	1	"Female does not begin incubation until penultimate egg is laid" as quoted from Martin and& Parrish (2000).
duration from start of incubation to hatch (days)	I	12	Synchronous hatch after 11–12 d of incubation (Martin and Parrish 2000).
duration from hatch to fledging of nestlings (days)	N	12	"Young capable of short flights at 9–10 d; leave nest typically at 11–12 d (Baepler 1968, Johnsgard 1979)" as quoted from Martin and Parrish (2000).

Parameter	Code	Value	Description
duration since nest failure due to other reasons until female initiates new nest (days)	We	7	"Renesting attempts are common (Baepler 1968)" as quoted from Martin and Parrish (2000). No specific data on duration of We. Assume renesting occurs rapidly after nest failure with a typical value of We of 7 d.
duration since successful fledging until female initiates new nest (days)	Wf	7	"Often produces a second clutch after the first clutch (Kaspari and Joern 1993)" as quoted from Martin and Parrish (2000). No specific data on duration of Wf. Assume renesting occurs rapidly after successful fledging with a typical value of Wf of 7 d.
female body weight (g) during breeding season	BdyWt	29	"In California, both sexes combined, mean 29.0 g ± 1.94 SD (range 24.7–33.3, $n = 49$; Dunning 1993). Mean adult C. g. strigatus, e. Great Basin and Colorado Plateau, UT: 28.3 g ± 2.24 SD (range 22.8–31.3, $n = 21$); Bonneville Basin, UT: adult males 28.3 g ± 1.29 SD (range 25.4–31.3, $n = 45$), adult females 27.8 g ± 2.58 SD (range 22.5–35.5, $n = 27$); hatch year 26.3 g ± 2.03 SD (range 20.0–31.9, $n = 167$; JWM). From Kansas (location unknown), breeding adult males 30.1 g (range 27.2–33.0, $n = 12$) and females 30.7 g (range 25.5–32.6, $n = 8$; Rising and Beadle 1996)" as quoted from Martin and Parrish (2000). Assume typical breeding season female weight of 29 g.
diet composition during breeding season		O	"Categorized as a ground-foraging omnivore during the breeding season, and a ground-gleaning granivore during the nonbreeding period (DeGraaf et al. 1985). In breeding season, eats more insects (biomass) than seeds (Kaspari and Joern 1993)" as quoted from Martin and Parrish (2000). Martin et al. (1951) show spring and summer adult diet of about 50% insects and 50% seeds. Juvenile diet: "Acidids, Lepidoptera larvae, and tettigonids were prevalent, with fewer adult Coleopterans, although proportions of prey items varied annually depending on availability (Kaspari and Joern 1993)" as quoted from Martin and Parrish (2000). Assume adults consume 50% insects and 50% seeds and juveniles 100% invertebrates.
mean number of nest attempts/ female/ season			
mean number of successful broods/ female/ season			"Normally one brood, two broods not uncommon" as quoted from Martin and Parrish (2000).
mean number of fledglings/ successful nest	fpsn	2.9	In OK, 14 of 33 nests successfully fledged 1 or more nestlings for a mean of 2.9 fledglings/successful nest (based on data from Table 1 in Newman 1970).
mean number of fledglings/ female/ season (ARS)	ARS		

Reference list

Baepler, D. H. 1968. Lark sparrow. Pages 886-902 *in* Life histories of North American cardinals, grosbeaks, buntings, towhees, finches, sparrows, and allies. (O. L. Austin, Ed.) *U.S. Nat. Mus. Bull.* 237.

Dunning Jr., J. B. 1993. Avian body masses. CRC Press, Boca Raton, FL.

Johnsgard, P. A. 1979. Birds of the Great Plains: breeding species and their distribution. Univ. of Nebraska Press, Lincoln.

Kaspari, M. and A. Joern. 1993. Prey choice by three insectivorous grassland birds: reevaluating opportunism. *Oikos* 68:414-430.

Martin, J. W. and J. R. Parrish. 2000. Lark Sparrow (*Chondestes grammacus*), The Birds of North America Online (A. Poole, Ed.). Ithaca: Cornell Lab of Ornithology; Retrieved from the Birds of North America Online: http://bna.birds.cornell.edu/bnaproxy.birds.cornell.edu/bna/species/488

Mcnair, D. B. 1985. A comparison of oology and nest record card data in evaluating the reproductive biology of Lark Sparrows, *Chondestes grammacus*. *Southwest. Nat.* 30:213-224.

Newman, G. A. 1970. Cowbird parasitism and nesting success of Lark Sparrows in southern Oklahoma. *Wilson Bull.* 82:304-309.

Rising, J. D. and D. Beadle. 1996. A guide to the identification and natural history of sparrows of the United States and Canada. Academic Press, New York.

Species Life-History Profiles – 12 December 2013

Lark bunting (*Calamospiza melanocorys*) — Four-letter Alpha Code: LARB

Species life-history parameters	Model Code	Typical value	Rationale
daily mortality rate during laying & incubation	m1	0.056	Based on the data from Table 3 in Jehle et al. (2004), the weighted mean of the Mayfield estimates for overall nest success during incubation at three sites in CO is 0.53 (using an 11 d incubation period), which translates to a daily nest mortality rate of 0.056 during incubation.
daily mortality rate during nestling-rearing	m2	0.091	Based on the data from Table 3 in Jehle et al. (2004), the weighted mean of the Mayfield estimates for overall nest success during nestling rearing at three sites in CO is 0.47 (using an 8 d nestling period), which translates to a daily nest mortality rate of 0.091 during nestling rearing.
date of first egg of first nest (dd-mmm)	T1	25-May	"Generally species probably not double-brooded owing to extensive early postbreeding migration (P. Creighton in Strong 1971, TGS). Egg-laying generally begins mid-May (Fig. 4). In 4-yr study in Weld Co., CO, earliest eggs 12 May; and latest clutch initiated 28 Jul (Porter and Ryder 1974).
date of first egg of last nest (dd-mmm)	Tlast	25-Jun	In Wallace Co., KS, egg dates ranged from 19 May to 12 Jul (TGS). Earliest egg dates in S. Dakota, 26 May (S. Dakota Ornithol. Union 1991); in Montana, 1 Jun (Whittle 1922). Mean/modal dates for egg-laying are lacking" as quoted from Shane (2000). Based on Figure 4 in Shane (2000), assume the core period for egg laying starts and ends on May 25 and June 25, respectively.
length of rapid follicle growth period (RFG) for each egg (days)	rfg	4	Based on allometric equation (RFG = 2.852*Egg mass ^0.31) from Alisauskas and Ankney (1992), using an egg mass of 3.17 g, assume rfg = 4 d.
mean clutch size	clutch	4	"In Colorado: 1968 mean, 3.6 (range 2–5, n = 43; Creighton 1971b); 1970 mean, 3.9 (range 2–6, n = 31); 1971 mean, 4.0 (range 2–5, n = 37; Strong 1971). In w. Kansas, mean clutch size, 21 May–3 Jun, 4.29 ± 0.488 SD (range 4–5, n = 7); 14 Jun–23 Jun, mean 4.67 ± 0.500 SD (range 4–5, n = 9); 24 Jun–12 Jul, mean 4.22 ± 0.441 SD (range 4–5, n = 9; TGS)" as quoted from Shane (2000). Assume typical clutch size of 4 eggs.
mean intra-egg laying interval (days)	eli	1	"In Colorado, 1 egg laid in early morning (before 05:30 MST) each day (Creighton 1971b)" as quoted from Shane (2000).
egg on which female typically begins incubation—penultimate (1) or last (0)	penult	1	"Incubation begins with penultimate egg (Pleszczynska 1977)" as quoted from Shane (2000).
duration from start of incubation to hatch (days)	I	12	"In Colorado, 11.7 d (n = 90 nests; Creighton and Baldwin 1974)" as quoted from Shane (2000).
duration from hatch to fledging of nestlings (days)	N	8	"Undisturbed young leave nest 8–9 d after hatching (Baldwin et al. 1969). Occasionally leave at 7 d (C. W. Huntley pers. comm.)" as quoted from Shane (2000).
duration since nest failure due to other reasons until female initiates new nest (days)	We	6	"In Kansas, commonly renests after a heavy nest loss during first attempts in May (TGS)" as quoted from Shane (2000). No specific data found on value for We, but assume that renesting after failure occurs rapidly and that a typical duration for We is 6 d.

143

duration since successful fledging until female initiates new nest (days)	Wf	100	Although Shane (2000) reports that "in Jun 1998, 2 pair of buntings started laying second brood 5–6 d after successful fledging of first nests in Weld Co., CO (B. Lyon pers. comm.)," lark buntings are not typically double brooded (P. Creighton in Strong 1971).
female body weight (g) during breeding season	BdyWt	37	Based on data from Table 4 of Shane (2000), female body weight during egg laying period is approximately 37 g.
diet composition during breeding season	O	O	"Summer: grasshoppers, weevils, ants, scarab beetles, true bugs, seeds of wild plants, grains, and some leafy matter (Martin et al. 1951, Baldwin et al. 1969). Summer: In Colorado, plant food averages 36–38% of diet, including seeds of oats (*Avena*), Indian ricegrass (*Oryzopsis*), smartweed (*Polygonum*), buffalo grass (*Buchloe*), sunflower (*Helianthus*), gromwell (*Lithospermum*), spiderwort (*Tradescantia*), triple-awn grass (*Aristida*), wheat (*Triticum*), pigweed (*Amaranthus*), Russian thistle (*Salsola*), verbena (*Verbena*), bulrush (*Scirpus*), sedge (*Carex*), and ball cactus (*Mamillaria*). Animal food averages 62–64% of diet; composed mostly of invertebrates (Acrididae, Curculionidae, Formicidae, Scarabaeidae, Tenebrionidae, Carabidae, Meloidae, Ichneumonidae, Cerambycidae, Anthomyiidae, Sphecidae, Chrysomelidae, Calliphoridae, and Cicadellidae; Baldwin et al. 1969, Baldwin 1973)." as quoted from Shane (2000). Assume typical adult diet of 64% invertebrates and 36% seeds, while juveniles consume 100% invertebrates.
mean number of nest attempts/ female/ season			
mean number of successful broods/ female/ season			"Normally only 1 brood reared/season (P. Creighton in Strong 1971)" as quoted from Shane (2000).
mean number of fledglings/ successful nest	fpsn	2.2	"Reproductive success for 3 studies in Weld Co., CO (number of fledglings/number of eggs), was 49.8%, averaging 2.5 fledglings/successful nest (Strong 1971); 40.9% success, or 2.6 fledglings/successful nest (Porter and Ryder 1974); and 39.1% success, or 1.6 fledglings/successful nest (Giezentanner and Ryder 1969)" as quoted from Shane (2000). The mean of these 3 estimates is 2.2 fledglings/successful nest.
mean number of fledglings/ female/ season (ARS)	ARS		

Reference list

Alisauskas, R. T., and C. D. Ankney. 1992. The cost of egg laying and its relationship to nutrient reserves in waterfowl. In: Ecology and Management of Breeding Waterfowl. B. Batt (Ed.), University of Minnesota Press. Pp 30-61.

Baldwin, P. H. 1973. The feeding regime of granivorous birds in shortgrass prairie in Colorado, in productivity, population dynamics and systematics of granivorous birds. Polish Sci. Publ., Warsaw.

Baldwin, P. H., J. D. Butterfield, P. D. Creighton, and R. Shook. 1969. Summer ecology of the Lark Bunting. Tech. Rep. no. 29. Grassland Biome, U.S. Int. Biol. Prog.

Creighton, P. D. 1971b. Nesting of the Lark Bunting in north-central Colorado. Tech. Rep. no. 68. Grassland Biome, U.S. Int. Biol. Prog.

Creighton, P. D. and P. H. Baldwin. 1974. Habitat exploitation by an avian ground-foraging guild. Tech. Rep. no. 263. Grassland Biome, U.S. Int. Biol. Prog.

Giezentanner, J. B. and R. A. Ryder. 1969. Avian distribution and population fluctuations at the Pawnee site. Tech. Rep. no. 28. Grassland Biome, U.S. Int. Biol. Prog.

Martin, A. C., H. S. Zim, and A. L. Nelson. 1951. American wildlife and plants. McGraw-Hill, New York.

Pleszczynska, W. 1977. Polygyny in the Lark Bunting. Ph.D. thesis. Univ. of Toronto, Toronto, ON.

Porter, D. K. and R. A. Ryder. 1974. Avian density and productivity studies and analysis on the Pawnee site in 1972. Tech. Rep. no. 252. Grassland Biome, U.S. Int. Biol. Prog.

Shane, T. G. 2000. Lark Bunting (*Calamospiza melanocorys*), The Birds of North America Online (A. Poole, Ed.). Ithaca: Cornell Lab of Ornithology; Retrieved from the Birds of North America Online: http://bna.birds.cornell.edu.bnaproxy.birds.cornell.edu/bna/species/542

South Dakota Ornithologists' Union. 1991. The birds of South Dakota. 2nd ed. S. Dakota Ornithol. Union, Aberdeen.

Strong, M. A. 1971. Avian productivity on the shortgrass prairie of northcentral Colorado. Master's Thesis. Colorado State Univ., Fort Collins.

Whittle, C. L. 1922. Miscellaneous bird notes from Montana. *Condor* 24:73-81.

Savannah sparrow (*Passerculus sandwichensis*)			Four-letter Alpha Code: SAVS
Species life-history parameters	Model Code	Typical value	Rationale
daily mortality rate during laying & incubation	m1	0.052	In MN and ND, the overall Mayfield nest success rate is 0.314 and the calculated Mayfield daily nest survival rate is 0.948 ± 0.003 (SE, n=687, Winter et al. 2004), so daily nest mortality rate is 0.052.
daily mortality rate during nestling-rearing	m2	0.052	In North Dakota, egg records are from May 28 to July 9. Minnesota egg records extend from May 30 to July 21 (Johnsgard 2009).
date of first egg of first nest (dd-mmm)	T1	May-28	
date of first egg of last nest (dd-mmm)	Tlast	Jul-21	
length of rapid follicle growth period (RFG) for each egg (days)	rfg	4	Based on allometric equation (RFG = 2.852*Egg mass ^0.31) from Alisauskas and Ankney (1992), using an egg mass of 2.2 g, assume rfg = 4 d.
mean clutch size	clutch	4	"Geographical variation. Range: 2–6 eggs, most commonly 4; Quebec, 4.0 (SD = 0.1, n = 214, 2–5; Bédard and Meunier 1983); New Brunswick, 4.0 (0.5, 284, 2–5; Dixon 1978); Sable Is., NS, 4.3 (0.1, 231, 2–6; Stobo and McLaren 1975); coastal Nova Scotia, 4.2 (24, 3–5; Welsh 1975); Minnesota and North Dakota, 4.1 (0.1, 687; Winter et al. 2004)" as quoted from Wheelwright and Rising (2008).
mean intra-egg laying interval (days)	eli	1	"One egg/d (Potter 1974), usually in early morning. Infrequently, 2 d between laying of successive eggs (NTW)" as quoted from Wheelwright and Rising (2008).
egg on which female typically begins incubation–penultimate (1) or last (0)	penult	1	"Incubation is infrequent and irregular until penultimate egg is laid (Stobo and McLaren 1975, Dixon 1978)" as quoted from Wheelwright and Rising (2008).
duration from start of incubation to hatch (days)	I	12	"Incubation period averages: Quebec, 12.0 d (Bédard and Meunier 1983); Kent Is., NB, 11.8 d (Dixon 1978); s. California, 13.2 d (Davis et al. 1984); Sable Is., NS, 12.5 d (9–15; Stobo and McLaren 1975; coast of Nova Scotia, 10 d (Welsh 1975)" as quoted from Wheelwright and Rising (2008).
duration from hatch to fledging of nestlings (days)	N	10	"Young remain in the nest for an average of 10.9 d (1.0, 111, 8–13) on Sable I., NS (Stobo and McLaren 1975); coastal Nova Scotia, 9.4 d (8–11, Welsh 1975); Kent I., NB, 9 d (Dixon 1978); Quebec, 9.8 d (7–10; Bédard and Meunier 1983, Bédard and LaPointe 1985); Michigan, 8 d (Potter 1974); Newfoundland, 9–11 d (Threlfall and Cannings 1979); California, 7–9 d (Williams and Nagy 1985). Disturbance after day 8 induces premature nestling departure" as quoted from Wheelwright and Rising (2008).
duration since nest failure due to other reasons until female initiates new nest (days)	We	5	Wheelwright and Schultz (1994) reported that clutch initiation after loss took 4.9 d for 1st year females and 3 d for older females. Assume typical duration for We is 5 d.

duration since successful fledging until female initiates new nest (days)	Wf	10	"Parents feed and remain with fledglings until they are at least 21 d old (Ross 1980a). Post-fledging care lasts a median of 15 d and does not differ significantly between male and female parents; some parents, especially females that do not lay a second clutch, remain with fledglings for > 25 d after fledging (Potter 1974, Stobo and McLaren 1975, Wheelwright et al. 2003). Given 12-d incubation period, 11-d nestling period, and 27–37-d interval between hatching of successive clutches (see above), female ordinarily provides only 4 to 14 d post-fledging parental care before constructing a new nest and completing second clutch" as quoted from Wheelwright and Rising (2008). Assume typical duration for Wf is 10 d.
female body weight (g) during breeding season	BdyWt	19.5	Based on data in Appendix 2 of Wheelwright and Rising (2008), the weighted mean of body weight for typical savannah sparrows is 19.5 g.
diet composition during breeding season		I	"Breeding season: mostly adult and larval insects, spiders, seeds and fruits, but occasionally insect eggs, millipedes, isopods, amphipods, decapods, mites, small mollusks (Martin et al. 1951, Baird 1968, Maher 1979, Meunier and Bédard 1984). Upon arrival at breeding grounds, insects are scarce and leaves of deciduous plants have not emerged, so diet initially consists mainly of seeds, then switches almost exclusively to insects for two months (Weatherhead 1979a, NTW). Nestlings fed almost entirely insects and spiders" as quoted from Wheelwright and Rising (2008). Assume that during peak of breeding season, adult females and juveniles consume 100% invertebrates.
mean number of nest attempts/ female/ season			
mean number of successful broods/ female/ season			
mean number of fledglings/ successful nest	fpsn	3.5	In MN and ND, the mean number of fledglings per successful nest is 3.50 ± 0.05 (SE, n=385, Winter et al. 2004)
mean number of fledglings/ female/ season (ARS)	ARS		

Reference list

Baird, J. 1968. *Passerculus sanwichensis savannah*, Eastern Savannah Sparrow. Pages 678-698 *in* Life Histories of North American cardinals, grosbeaks, buntings, towhees, finches, sparrows and allies. Vol. Bull. U.S. Natl. Mus. (O. L. Austin Jr., Ed.)

Bedard, J. and G. Lapointe. 1985. Influence of parental age and season on Savannah Sparrow reproductive success. *Condor* 87(1):106-110.

Bedard, J. and M. Meunier. 1983. Parental care in the Savannah Sparrow. *Can. J. Zool.* 61(12):2836-2843.

Davis, S. D., J. B. Williams, W. J. Adams, and S. L. Brown. 1984. The effect of egg temperature on attentiveness in the Belding's Savannah Sparrow. *Auk* 101(3):556-566.

Dixon, C. L. 1978. Breeding biology of the Savannah Sparrow on Kent Island. *Auk* 95(2):235-246.

Johnsgard, P. A., 2009. Birds of the Great Plains: Family Fringillidae (Grosbeaks, Finches, Sparrows, and Buntings). Retrieved from Birds of the Great Plains (Revised edition 2009) by Paul Johnsgard. Paper 60. http://digitalcommons.unl.edu/bioscibirdsgreatplains/60.

Maher, W. J. 1979. Nestling diets of prairie passerine birds at Matador, Saskatchewan, Canada. *Ibis* 121(4):437-452.

Martin, A. C., H. S. Zim, and A. L. Nelson. 1951. American wildlife and plants. Pages ix+500p. Illus. U.S. Dept. of Interior, Washington, DC.

Meunier, M. and J. Bedard. 1984. Nestling foods of the Savannah Sparrow. *Can. J. Zool.* 62(1):23-27.

Potter, P. E. 1974. Breeding behavior of Savannah Sparrows in southeastern Michigan. *Jack-Pine Warbler* 52(2):50-63.

Ross, H. A. 1980a. The reproductive rates of yearloing and older Ipswich Sparrows, *Passerculus sandwichensis pinceps*. Can. J. Zool.-Rev. Can. Zool. 58(9):1557-1563.

Stobo, W. T. and I. A. Mc Laren. 1971. Late winter distribution of the Ipswich Sparrow. *Amer. Birds* 25(6):941-944.

Stobo, W. T. and I. A. McLaren. 1975. The Ipswich sparrow. *Proc. Nova Scot. Inst. Sci.* 27:1-105.

Threlfall, W. and R. J. Cannings. 1979. Growth of nestling Savannah Sparrows. *Bird-Banding* 50(2):164-166.

Welsh, D. A. 1975. Savannah Sparrow breeding and territoriality on a Nova Scotia dune beach. *Auk* 92(2):235-251.

Wheelwright, N. T. and J. D. Rising. 2008. Savannah Sparrow (*Passerculus sandwichensis*), The Birds of North America Online (A. Poole, Ed.). Ithaca: Cornell Lab of Ornithology; Retrieved from the Birds of North America Online: http://bna.birds.cornell.edu.bnaproxy.birds.cornell.edu/bna/species/045.

Wheelwright, N. T. and C. B. Schultz. 1994. Age and reproduction in Savannah Sparrows and Tree Swallows. *J. An. Ecol.* 63(3):686-702.

Wheelwright, N. T., K. A. Tice, and C. R. Freeman-Gallant. 2003. Postfledging parental care in Savannah sparrows: sex, size and survival. *An. Behav.* 65:435-443.

Williams, J. B. and K. A. Nagy. 1985. Water flux and energetics of nestling Savannah Sparrows in the field. *Physiol. Zool.* 58(5):515-525.

Winter, M., D. H. Johnson, J. A. Shaffer, and W. D. Svedarsky. 2004. Nesting biology of three grassland passerines in the northern tallgrass prairie. *Wilson Bull.* 116(3):211-223.

Grasshopper sparrow (*Ammodramus savannarum*) — Four-letter Alpha Code: GRSP

Species life-history parameters	Model Code	Typical value	Rationale
daily mortality rate during laying & incubation	m1	0.0428	"Nesting success, defined as producing at least 1 fledgling/nest, varies considerably throughout species' range. The few published studies provide no clear regional trends as yet. Moderate levels of success reported from Maine (40–50%; Vickery et al. 1992a) and se. Nebraska (52%; Delisle and Savidge 1996), but lower success rates noted in Florida (<25% in 1993; W. G. Shriver, PDV and Illinois (<35%; J. R. Herkert pers. comm.), due primarily to nest predation. In Iowa, nests (n = 41) in grassed strips in agricultural land had high failure rates (84%), of which 80% was due to predation, 2% to agricultural practices, 2% to brood parasitism; these nests produced 0.8 fledglings/yr (N. S. Basore and L. B. Best unpubl. data)" as quoted from Vickery (1996). If a typical nest success rate is 35%, over a 24 d nest period (i.e., 9+12+4-1) the daily nest mortality rate is 0.0428.
daily mortality rate during nestling-rearing	m2	0.0428	
date of first egg of first nest (dd-mmm)	T1	May-11	"In general, breeding season protracted; depending on favorable weather, species can produce ≥ 2 broods annually, even in northern portion of range (Vickery et al. 1992c, J. R. Herkert pers. comm.). In Oklahoma, 2 and sometimes 3 broods; late nest found mid-Aug (Easley 1983). In Pennsylvania, 2 broods annually (Smith 1963), but species becomes more numerous and widespread in Jun; such late-arriving breeders likely produce just 1 brood/season (Santner 1992); Jun arrivals are likely first-summer breeders, with experienced adults arriving earlier and potentially raising 2 clutches (Wiens 1969). In Alabama, eggs 11 May–15 Jul (n = 16 nests (Imhof 1976)" as quoted from Vickery (1996).
date of first egg of last nest (dd-mmm)	Tlast	Jul-15	
length of rapid follicle growth period (RFG) for each egg (days)	rfg	4	RFG for passerines typically 3-4 d.
mean clutch size	clutch	4	"Clutch size rangewide: 4.30 ± 0.69 (SD, n =438; McNair 1987); second clutches generally smaller, often with 3 eggs" as quoted from Vickery (1996).
mean intra-egg laying interval (days)	eli	1	No specific information, but assume egg-laying interval of 1 d like other passerines.
egg on which female typically begins incubation–penultimate (1) or last (0)	penult	1	"Uncertain, but persistent incubation probably starts with penultimate egg of clutch" as quoted from Vickery (1996).
duration from start of incubation to hatch (days)	I	12	"11–13 d for eastern *A. s. pratensis*, *A. s. floridanus*, and *A. s. perpallidus* (Nicholson 1936, Smith 1968, PDV). Probably similar for *A. s. ammolegus* and Caribbean and Central and South American subspecies" as quoted from Vickery (1996).
duration from hatch to fledging of nestlings (days)	N	9	"Nestlings generally remain in nest 8–9 d, 9 d in Michigan and Pennsylvania (Walkinshaw 1940, Smith 1963), 6–8 d in Nebraska (Kaspari and O'Leary 1988]" as quoted from Vickery (1996).
duration since nest failure due to other reasons until female initiates new nest (days)	We	6	No specific data found on value for We, but assume renesting after failure occurs rapidly. Assume typical value of We of 6 d.

duration since successful fledging until female initiates new nest (days)	Wf	10	"Fledglings disperse immediately from vicinity of nest. Not known whether dispersal is independent or promoted by parents. Both male and female provide postfledging care for young; duration of parental care not known. Based on 12-d incubation period, 9-d nestling period, and 25- to 40-d interval between hatching successive clutches, female ordinarily provides only 4–19 d postfledging parental care before she initiates nest construction for second clutch" as quoted from Vickery (1996). Assume that a typical value for Wf is 10 d.
female body weight (g) during breeding season	BdyWt	18	"Mean mass (g) in breeding season: $A.\ s.\ floridanus$: male ($n = 25$) 17.17 ± 0.2 (SE), female ($n = 5$) 18.38 ± 0.39 (Delany et al. 1994). $A.\ s.\ pratensis$ in Connecticut: male ($n = 8$) 17.82 ± 0.26 (SE), female ($n = 1$) 18.75 (Crossman 1989). In Maine, male ($n = 42$) 17.34 ± 0.17 (SE, range 14.5–20; J. V. Wells pers. comm.). $A.\ s.\ perpallidus$: male ($n = 16$) 16.8 ± 1.3 (SD), female ($n = 6$) 17.0 ± 0.7 (SD) (Collier 1994)" as quoted from Vickery (1996). Assume typical breeding season female weight of 18 g.
diet composition during breeding season		O	"Insects, primarily grasshoppers, form major part of summer diet throughout breeding range. Summer diet 61% invertebrates, 39% seeds ($n = 100$). In Wisconsin, young fed, in order of frequency (number of observed food items), Lepidoptera larvae 29, Odonata (Zygoptera) 4, Hemiptera 1, Homoptera 1, Coleoptera larvae 1, Diptera 1, Arachnida 1, Oligochaete 1 (Wiens 1969)" as quoted from Vickery (1996).
mean number of nest attempts/ female/ season			"Because predation rates are often >50%, Grasshopper Sparrows frequently renest. Number of renesting attempts not recorded, probably 3–4 annually" as quoted from Vickery (1996).
mean number of successful broods/ female/ season			
mean number of fledglings/ successful nest	fpsn	3.2	No quantitative information found. Fledging is difficult to document because of ground nests from which fledglings walk. In FL, Perkins et al. (2003) estimated 3.2 fledglings/successful nest
mean number of fledglings/ female/ season (ARS)	ARS		In FL, Perkins et al. (2003) from estimated annual productivity per pair to be between 1.46 and 4.09 for all sites and all years.

Reference list

Collier, C. L. 1994. Habitat selection and reproductive success of the Grasshopper Sparrow at the Santa Rosa Plateau Ecological Reserve. Master's Thesis. San Diego State Univ. San Diego, CA.

Crossman, T. I. 1989. Habitat use by Grasshopper and Savannah sparrows at Bradley International Airport and management recommendations. Master's Thesis. Univ. of Connecticut, Storrs.

Delany, M. F., C. T. Moore, and D. R. Progulske, Jr. 1994. Distinguishing gender of Florida Grasshopper Sparrows using body measurements. $Fla.\ Field\ Nat.$ 22:48-51.

Delisle, J. M. and J. A. Savidge. 1996. Reproductive success of Grasshopper Sparrows in relation to distance from Conservation Reserve Program field edges. *Prairie Natural.* 28(3):107-114.

Easley, W. 1983. Late nesting of Grasshopper Sparrow in Texas County, Oklahoma. *Bull. Okla. Ornithol. Soc.* 16:23-24.

Imhof, T. A. 1976. Alabama birds. Univ. of Alabama Press, Tuscaloosa.

Kaspari, M. and H. O'Leary. 1988. Nonparental attendants in a north-temperate migrant. *Auk* 105:792-793.

Mcnair, D. B. 1987. Egg data slips-are they useful for information on egg-laying dates and clutch size? *Condor* 89:369-376.

Nicholson, W. H. 1936. Notes on the habits of the Florida Grasshopper Sparrow. *Auk* 53:318-319.

Perkins, D. W., P. D. Vickery, and W. G. Shriver. 2003. Spatial dynamics of source-sink habitats: Effects on rare grassland birds. *J. Wildl. Manage.* 67:588-599.

Santner, S. 1992. Grasshopper Sparrow. Pages 384-385 *in* Atlas of breeding birds in Pennsylvania. (Brauning, D. W., Ed.) Univ. of Pittsburgh Press, Pittsburgh.

Smith, R. L. 1963. Some ecological notes on the Grasshopper Sparrow. *Wilson Bull.* 75:159-165.

Smith, R. L. 1968. Grasshopper Sparrow. Pages 725-745 *in* Life histories of North American cardinals, grosbeaks, buntings, towhees, finches, sparrows, and allies. Pt. 2. (O. L. Austin Jr., Ed.) *U.S. Natl. Mus. Bull.* 237.

Vickery, P. D. 1996. Grasshopper Sparrow (*Ammodramus savannarum*), The Birds of North America Online (A. Poole, Ed.). Ithaca: Cornell Lab of Ornithology; Retrieved from the Birds of North America Online: http://bna.birds.cornell.edu.bnaproxy.birds.cornell.edu/bna/species/239

Vickery, P. D., M. L. Hunter, Jr., and J. V. Wells. 1992a. Use of a new reproductive index to evaluate relationship between habitat quality and breeding success. *Auk* 109:697-705.

Vickery, P. D., M. L. Hunter, Jr., and J. V. Wells. 1992c. Evidence of incidental nest predation and its effects on nests of threatened grassland birds. *Oikos* 63:281-288.

Walkinshaw, L. H. 1940. Some Michigan notes on the Grasshopper Sparrow. *Jack-Pine Warbler* 18:50-59.

Wiens, J. A. 1969. An approach to the study of ecological relationships among grassland birds. *Ornithol. Monogr.* no. 8.

Nuttall's white-crowned sparrow (*Zonotrichia leucophrys nuttali*)[1]

Four-letter Alpha Code: NWCS

Species life-history parameters	Model Code	Typical value	Rationale
daily mortality rate during laying & incubation	m1	0.0197	At 2 sites in CA, daily nest mortality rate of 0.0197 during 12 d incubation phase based on 6-yr study of *Z. l. nuttalli* (Petrinovich and Patterson 1983).
daily mortality rate during nestling-rearing	m2	0.0766	At 2 sites in CA, daily nest mortality rate of 0.0766 during 10 d nestling rearing phase based on 6-yr study of *Z. l. nuttalli* (Petrinovich and Patterson 1983).
date of first egg of first nest (dd-mmm)	T1	20-Mar	In CA, *Z. l. nuttalli* begins laying mid-March to early April and have the longest breeding season (about 100 d) among all subspecies. Blanchard (1941) shows that in most years few nest initiated prior to March 20. Assume egglaying initiated between March 20 and Jul 7.
date of first egg of last nest (dd-mmm)	Tlast	7-Jul	
length of rapid follicle growth period (RFG) for each egg (days)	rfg	4	Based on allometric equation (RFG = 2.852*Egg mass ^0.31) from Alisauskas and Ankney (1992), using and egg mass of 3.04 g, assume rfg = 4 d.
mean clutch size	clutch	3	*Z. l. nuttalli*: mean 3.15 (SD=0.70, n=1062; Petrinovich and Patterson 1983) and 3.27 (n=215).
mean intra-egg laying interval (days)	eli	1	Typically 1 egg/d (Chilton et al. 1995).
egg on which female typically begins incubation–penultimate (1) or last (0)	penult	1	"Consistent daytime incubation with nighttime sitting doe not usually begin until penultimate egg is laid" (Chilton et al. 1995).
duration from start of incubation to hatch (days)	I	12	"Generally averages 12 d among all subspecies, but variable within populations and years" (Chilton et al. 1995).
duration from hatch to fledging of nestlings (days)	N	10	Fledglings depart nest at 8.5 to 10 d of age (Chilton et al. 1995). Assume fledging typically occurs at 10 d.
duration since nest failure due to other reasons until female initiates new nest (days)	We	7	First egg in new nest after failure in mean of 6.8 ± 3.9 (n=26, range 2-16) for *Z. l. oriantha* (Morton et al. (1972).
duration since successful fledging until female initiates new nest (days)	Wf	20	For *Z. l. nuttalli* "first egg of second brood is laid 20 d (n=6, range=14-29) after fledging first brood. For 1 pair that fledged 3 broods, first interbrood interval was 14 d, second was 20 d (Blanchard 1941)" of as cited in Chilton et al. (1995).
female body weight (g) during breeding season	BdyWt	32	32.0 ± 2.18 (n=50) for *Z. l. nuttalli* (Dunning 1984).
diet composition during breeding season		O	During fall and winter adults consume primarily seeds, but during breeding season the proportion of insects increase significantly (Chilton et al. 1995). Martin et al. (1951) estimates breeding season diet contains 64% plants (predominantly weed seeds) and 36% invertebrates.

mean number of nest attempts/ female/ season		3.06	The weighted mean of 3.06 nests/pair based on 174 pairs at Twin Peaks and 86 pairs at Presidio in CA (Petrinovich and Patterson 1983).
mean number of successful broods/ female/ season			
mean number of fledglings/ successful nest	fpsn	2.46	Estimate of 2.46 fledglings/successful nest based on 418 successful nests of *Z. l. nuttalli* in CA (Petrinovich and Patterson 1983).
mean number of fledglings/ female/ season (ARS)	ARS	2.48	Calculated ARS of 2.48 fledglings/female/year for *Z. l. nuttalli* in CA (Petrinovich and Patterson 1983).

[1] Where possible parameters reflect information on the Nuttall's white-crowned sparrow (*Zonotrichia leucophrys nuttalli*), but may include information on the Puget Sound white-crowned sparrow (*Z. l. pugetensis*) because these subspecies are most closely associated with agriculture during breeding. Other subspecies typically breed outside of agricultural areas. The mountain white-crowned sparrow (*Z. l. oriantha*) breeds in high elevation habitats of western U.S. and Canada. The Gambel's white-crowned sparrow (*Z. l. gambelii*) breeds in Alaska and western Canada. The eastern white-crowned sparrow (*Z. l. leucophrys*) breeds in the arctic region of eastern Canada, but winters in, and migrates through, eastern U. S.

Reference list

Alisauskas, R. T., and C. D. Ankney. 1992. The cost of egg laying and its relationship to nutrient reserves in waterfowl. In: *Ecology and Management of Breeding Waterfowl.* B. Batt (Ed.). University of Minnesota Press. Pp 30-61.

Blanchard, B. D. 1941. The White-crowned Sparrows (*Zonotrichia leucophrys*) of the Pacific seaboard: environment and annual cycle. *Univ. Calif. Publ. Zool.* 46:1-178.

Chilton, G., M. C. Baker, C. D. Barrentine, and M. A. Cunningham. 1995. White-crowned Sparrow (*Zonotrichia leucophrys*). The Birds of North America Online (A. Poole, Ed.). Ithaca: Cornell Lab of Ornithology; Retrieved from the Birds of North America Online: http://bna.birds.cornell.edu.bnaproxy.birds.cornell.edu/bna/species/183.

DeWolfe, B. B. 1968. *Zonotrichia leucophrys nuttalli* (Ridgway), Nuttall's White-crowned Sparrow. Pages 1292-1324 *in* Life histories of North American cardinals, grosbeaks, buntings, towhees, finches, sparrows, and allies. (O. L. Austin, Ed.), *U.S. Natl. Mus. Bull.* 237, pt. 3.

Dunning, J. B. Jr. 1984. Body weights of 686 species of North American birds., Eldon Publishing, Cave Creek, AZ.

Morton, M. L., J. L. Horstmann, and J. M. Osborn. 1972. Reproductive cycle and nesting success of the Mountain White-crowned Sparrow (*Zonotrichia leucophrys oriantha*) in the central Sierra Nevada. *Condor* 74:152-163.

Petrinovich, L. and T. L. Patterson. 1983. The White-crowned Sparrow: reproductive success (1975-1980). *Auk* 100:811-825.

Puget Sound white-crowned sparrow (*Zonotrichia leucophrys pugetensis*) [1]		Four-letter Alpha Code: PSWS	
Species life-history parameters	Model Code	Typical value	Rationale
daily mortality rate during laying & incubation	m1	0.0197	At 2 sites in CA, daily nest mortality rate of 0.0197 during 12 d incubation phase based on 6-yr study of *Z. l. nuttalli* (Petrinovich and Patterson 1983).
daily mortality rate during nestling-rearing	m2	0.0766	At 2 sites in CA, daily nest mortality rate of 0.0766 during 10 d nestling rearing phase based on 6-yr study of *Z. l. nuttalli* (Petrinovich and Patterson 1983).
date of first egg of first nest (dd-mmm)	T1	25-Apr	In WA, *Z. l. pugetensis* begins laying first clutch in late April or early May and may initiate a second clutch during the first week of June and a third during the first week of July (Chilton et al. 1995). Assume egglaying starts April 25 and ends July 7.
date of first egg of last nest (dd-mmm)	Tlast	7-Jul	
length of rapid follicle growth period (RFG) for each egg (days)	rfg	4	Based on allometric equation (RFG = $2.852 \times \text{Egg mass}^{0.31}$) from Alisauskas and Ankney (1992), using and egg mass of 3.04 g, assume rfg = 4 d.
mean clutch size	clutch	4	*Z. l. pugetensis*: mean 3.84 (SD=0.83, n=23; Lewis 1975) and 4.04 (n=48) (DeWolf 1968).
mean intra-egg laying interval (days)	eli	1	Typically 1 egg/d (Chilton et al. 1995).
egg on which female typically begins incubation—penultimate (1) or last (0)	penult	1	"Consistent daytime incubation with nighttime sitting doe not usually begin until penultimate egg is laid" (Chilton et al. 1995).
duration from start of incubation to hatch (days)	I	12	"Generally averages 12 d among all subspecies, but variable within populations and years" (Chilton et al. 1995).
duration from hatch to fledging of nestlings (days)	N	10	Fledglings depart nest at 8.5 to 10 d of age (Chilton et al. 1995). Assume fledging typically occurs at 10 d.
duration since nest failure due to other reasons until female initiates new nest (days)	We	7	First egg in new nest after failure in mean of 6.8 ± 3.9 d (n=26, range 2-16) for *Z l. oriantha* (Morton et al. (1972).
duration since successful fledging until female initiates new nest (days)	Wf	9	"First egg of second brood is laid 8.9 d (n=5, range =6-15) after fledging first brood (Blanchard 1941)" for *Z. l. pugetensis* as cited in Chilton et al. (1995). For 1 pair that fledged 3 broods, first interbrood interval was 14 d, second was 20 d (Blanchard 1941)" of as cited in Chilton et al. (1995).
female body weight (g) during breeding season	BdyWt	25.3	25.3 ± 1.78 (n=50) for *Z l. pugetensis* (Dunning 1984).
diet composition during breeding season		O	During fall and winter adults consume primarily seeds, but during breeding season the proportion of insects increase significantly (Chilton et al. 1995). Martin et al. (1951) estimates breeding season diet contains 64% plants (predominantly weed seeds) and 36% invertebrates.

Species Life-History Profiles – 12 December 2013

mean number of nest attempts/ female/ season		
mean number of successful broods/ female/ season		
mean number of fledglings/ successful nest	fpsn	2.46
mean number of fledglings/ female/ season (ARS)	ARS	

Estimate of 2.46 fledglings/successful nest based on 418 successful nests of *Z. l. nuttalli* in CA (Petrinovich and Patterson 1983).

¹ Where possible parameters reflect the Puget Sound white-crowned sparrow (*Zonotrichia leucophrys pugetensis*), though some parameters are best described for the Nuttall's white-crowned sparrow (*Z. l. nuttalli*). These two subspecies are most closely associated with agriculture during the breeding season, while other subspecies tend to breeding outside of agricultural areas. The mountain white-crowned sparrow (*Z. l. oriantha*) breeds in high elevation habitats of western U.S. and Canada. The Gambel's white-crowned sparrow (*Z. l. gambelii*) breeds in Alaska and western Canada. The eastern white-crowned sparrow (*Z. l. leucophrys*) breeds in the arctic region of eastern Canada, but winters in, and migrates through, eastern U.S.

Reference list

Alisauskas, R. T., and C. D. Ankney. 1992. The cost of egg laying and its relationship to nutrient reserves in waterfowl. In: *Ecology and Management of Breeding Waterfowl*. B. Batt (Ed.), University of Minnesota Press. Pp 30-61.

Blanchard, B. D. 1941. The White-crowned Sparrows (*Zonotrichia leucophrys*) of the Pacific seaboard: environment and annual cycle. *Univ. Calif. Publ. Zool.* 46:1-178.

Chilton, G., M. C. Baker, C. D. Barrentine, and M. A. Cunningham. 1995. White-crowned Sparrow (Zonotrichia leucophrys), The Birds of North America Online (A. Poole, Ed.). Ithaca: Cornell Lab of Ornithology; Retrieved from the Birds of North America Online: http://bna.birds.cornell.edu/bna/species/183.

DeWolfe, B. B. 1968. *Zonotrichia leucophrys nuttalli* (Ridgway), Nuttall's White-crowned Sparrow. Pages 1292-1324 *in* Life histories of North American cardinals, grosbeaks, buntings, towhees, finches, sparrows, and allies. (O. L.Austin, Ed.), *U.S. Natl. Mus. Bull.* 237, pt. 3.

Dunning, J. B. Jr. 1984. Body weights of 686 species of North American birds., Eldon Publishing, Cave Creek, AZ.

Morton, M. L., J. L. Horstmann, and J. M. Osborn. 1972. Reproductive cycle and nesting success of the Mountain White-crowned Sparrow (*Zonotrichia leucophrys oriantha*) in the central Sierra Nevada. *Condor* 74:152-163.

Petrinovich, L. and T. L. Patterson. 1983. The White-crowned Sparrow: reproductive success (1975-1980). *Auk* 100:811-825.

Dark-eyed junco (*Junco hyemalis*)	Four-letter Alpha Code: DEJU		
Species life-history parameters	Model Code	Typical value	Rationale
daily mortality rate during laying & incubation	m1	0.0211	In VA, "Of 170 nests on territories of 93 males (1989–1993), 29.4% failed before hatching, additional 15.9% between hatching and fledging (Ketterson et al. 1996)" as quoted from Nolan et al. (2002). This translates to an overall apparent nest success rate of 55% (93/170). Over a 28 d nesting period (i.e., 12+13+4+1), this translates to a daily nest mortality rate of 0.0211.
daily mortality rate during nestling-rearing	m2	0.0211	
date of first egg of first nest (dd-mmm)	T1	27-Apr	Based on the MLBS, VA profile in Figure 3 of Nolan et al. (2002), the core egg laying period starts April 27 until June 30. The core "egg" bar extends to mid-July, but appears to represent eggs in nest, rather than initiation of egg laying, so period shorted by approximately 2 weeks.
date of first egg of last nest (dd-mmm)	Tlast	30-Jun	
length of rapid follicle growth period (RFG) for each egg (days)	rfg	4	Based on allometric equation (RFG = $2.852 \times \text{Egg mass}^{0.31}$) from Alisauskas and Ankney (1992), using and egg mass of 2.53 g, assume rfg = 4 d.
mean clutch size	clutch	4	"Modal clutch size 4, with 5- and 3-egg clutches not uncommon, latter especially near end of season" as quoted from Nolan et al. (2002).
mean intra-egg laying interval (days)	eli	1	Rate is 1 egg/d (Nolan et al. 2002).
egg on which female typically begins incubation–penultimate (1) or last (0)	penult	1	"Incubation usually begins late in day on which penultimate egg laid (Hostetter 1961, Sprunt 1968)" as quoted from Nolan et al. (2002).
duration from start of incubation to hatch (days)	I	13	"Incubation period (measured in various ways [or unspecified] in literature) reported as 12–13 d (11 d in 2 cases [Peck 1998]), with only slight variation among subspecies (Dixon 1924, DeGroot 1934, Eaton 1968, Phelps 1968c, Sprunt 1968, Whitney 1968)" as quoted from Nolan et al. (2002).
duration from hatch to fledging of nestlings (days)	N	12	"May fledge at age 10–11 d (hatching day = 1) if disturbed (i.e., total 11–12 d in nest); beginning day 12, usually burst from nest, try to run and/or fly; normally do not leave nest until day 12 or 13 (Hostetter 1961, Thatcher 1968, Weydemeyer 1971)" as quoted from Nolan et al. (2002).
duration since nest failure due to other reasons until female initiates new nest (days)	We	7	"Following nest failure, female builds new nest and replaces clutch, often several times (Smith and Andersen 1982)" as quoted from Nolan et al. (2002). Ketterson et al. (1992) reported We = 7.1 d (0.66 SE, N=15) for control pairs where investigator caused nest "failure" when nestlings were removed at 10 d old to be hand reared. Pairs were followed to determine date of first egg after "failure."
duration since successful fledging until female initiates new nest (days)	Wf	16	"Fourteen days after nest departure, at age 25–26 d, fledglings regularly feed selves; flight appears as maneuverable as in adults. Wolf et al. (1988) and subsequent MLBS studies, therefore, treated this as age of independence, although feeding occasionally observed 16 d after fledging (MLBS). Interval between fledging of first brood and laying of egg 1 in second-brood nests (1983–1986; Wolf et al. 1991), 15.9 d ± 1.1 SE (*n* = 23)" as quoted from Nolan et al. (2002).

female body weight (g) during breeding season	BdyWt	18.8	"Mean mass (g ± SD) of selected subspecies (from Dunning 1993 unless otherwise indicated; for some samples, details associated with variation are described): *J. h. hyemalis* (Pennsylvania), male 20.4 ± 1.21 (extremes 14.3–26.7, *n* = 2,819), female 18.8 ± 0.78 (extremes 14.3–25.1, *n* = 1,316; for seasonal variation in this subspecies, see below). *J. h. mearnsi* (Arizona; sexes pooled), 18.2 ± 1.30 (extremes 15.5–23.5, *n* = 221). *J. h. dorsalis* (Arizona; sexes pooled), 21.8 ± 1.40 (extremes 18.0–26.0, *n* = 170). *J. h. caniceps* (Arizona; sexes pooled), 19.6 ± 1.10 (extremes 18.0–23.0, *n* = 40)" as quoted from Nolan et al. (2002). Assume typical wt is 18.8 g.
diet composition during breeding season		O	Based on information in Martin et al (1951), junco consume 60% invertebrates and 40% plant material (mainly seeds) in April and May, but proportion plants in the diet jumps in June and thereafter. "*Food of Young, Kinds and Size of Items.* Food mostly arthropods (insects, spiders), occasionally plant material (*J. h. oreganus* [Phelps 1968a], *J. h. caniceps* [Thatcher 1968], *J. h. carolinensis* [millet from investigators' bait piles]) and grit (Hostetter 1968)" as quoted from Nolan et al. (2002). Assume adult diet is 60% invertebrates and 40% seeds, while juveniles consume 100% seeds.
mean number of nest attempts/ female/ season			
mean number of successful broods/ female/ season			"Varies geographically. Assuming female's early nests escape predation, 1 successful brood in north, 2 broods in south" as quoted from Nolan et al. (2002). In VA, Raouf et al. (1997) reported 1.17 successful nests per male.
mean number of fledglings/ successful nest	fpsn	3.08	"In nests that produced ≥1 fledgling, number of fledglings was 3.08 ± 0.112 SE" as quoted from Nolan et al. (2002).
mean number of fledglings/ female/ season (ARS)	ARS	3.9	In VA, Raouf et al. (1997) reported 3.9 fledglings/control male based on genetic fingerprinting. The study compared control and testosterone-treated males to examine role of extra-pair fertilizations on productivity of males. Although males have one social partner, extra-pair fertilizations are observed. For a population the mean for each female is likely similar to the mean for males.

Reference List

Alisauskas, R. T., and C. D. Ankney. 1992. The cost of egg laying and its relationship to nutrient reserves in waterfowl. In: *Ecology and Management of Breeding Waterfowl.* B. Batt (Ed.), University of Minnesota Press. Pp 30-61.

Degroot, D. S. 1934. Field observations from Echo Lake, California. *Condor* 36:6-9.

Dixon, J. 1924. Early nesting of the junco on the Berkeley campus. *Condor* 26:197.

Dunning, Jr., J. B. 1993. CRC handbook of avian body masses. CRC Press, Inc. Boca Raton, FL.

Eaton, S. W. 1968. Northern Slate-colored Junco. Pages 1029-1043 *in* Life histories of North American cardinals, grosbeaks, buntings, towhees, finches, sparrows, and their allies. Pt. 2. (O. L. Austin, Ed.), *U.S. Natl. Mus. Bull.* 237.

Hostetter, D. R. 1961. Life history of the Carolina Junco *Junco hyemalis carolinensis* Brewster. *Raven* 32:97-170.

Ketterson, E. D., V. Nolan, Jr., L. Wolf, and C. Ziegenfus. 1992. Testosterone and avian life histories: Effects of experimentally elevated testosterone on behavior and correlates of fitness in the Dark-eyed Junco (*Junco hyemalis*). *Am. Nat.* 140:980-999.

Ketterson, E. D., V. Nolan, Jr., M. J. Cawthorn, P. G. Parker, and C. Ziegenfus. 1996. Phenotypic engineering: using hormones to explore the mechanistic and functional bases of phenotypic variation in nature. *Ibis* 138:1-17.

Martin, A. C., H. S. Zim, and A. L. Nelson. 1951. American wildlife and plants. McGraw-Hill, New York.

Nolan, Jr., V., E. D. Ketterson, D. A. Cristol, C. M. Rogers, E. D. Clotfelter, R. C. Titus, S. J. Schoech and E. Snajdr. 2002. Dark-eyed Junco (Junco hyemalis). The Birds of North America Online (A. Poole, Ed.). Ithaca: Cornell Lab of Ornithology; Retrieved from the Birds of North America Online: http://bna.birds.cornell.edu/bnaproxy.birds.cornell.edu/bna/species/716

Peck, G. K. 1998. Ontario birds. *J. Ontario Field Nat.* 16:117.

Phelps, Jr., J. H. 1968a. Oregon Junco. Pages 1050-1071 *in* Life histories of North American cardinals, grosbeaks, buntings, towhees, finches, sparrows, and their allies. Pt. 2. (O. L. Austin, Ed.), *U.S. Natl. Mus. Bull.* 237.

Phelps, Jr., J. H. 1968c. Point Pinos Oregon Junco. Pages 1083-1089 *in* Life histories of North American cardinals, grosbeaks, buntings, towhees, finches, sparrows, and their allies. Pt. 2. (O. L. Austin, Ed.), *U.S. Natl. Mus. Bull.* 237.

Raouf, S. A., P. G. Parker, E. D. Ketterson, V. Nolan, Jr., and C. Ziegenfus. 1997. Testosterone affects reproductive success by influencing extra-pair fertilizations in male Dark-eyed Juncos, *Junco hyemalis*. *Proc. R. Soc. Lond. B.* 264:1599-1603.

Smith, K. G. and D. C. Andersen. 1982. Food, predation, and reproductive ecology of the Dark-eyed Junco in northern Utah. *Auk* 99:650-661.

Sprunt, Jr., A. 1968. Carolina Slate-colored Junco. Pages 1043-1049 *in* Life histories of North American cardinals, grosbeaks, buntings, towhees, finches, sparrows, and their allies. Pt. 2. (O. L. Austin, Ed.), *U.S. Natl. Mus. Bull.* 237.

Thatcher, D. M. 1968. Gray-headed Junco. Pages 1098-1126 *in* Life histories of North American cardinals, grosbeaks, buntings, towhees, finches, sparrows, and their allies. Pt. 2. (O. L. Austin, Ed.), *U.S. Natl. Mus. Bull.* 237.

Weydemeyer, W. 1971. Nesting habits of the Oregon Junco in Montana. *Wilson Bull.* 83:103-104.

Whitney, N. R. 1968. White-winged Junco. Pages 1021-1029 *in* Life histories of North American cardinals, grosbeaks, buntings, towhees, finches, sparrows, and their allies. Pt. 2. (O. L. Austin, Ed.), *U.S. Natl. Mus. Bull.* 237.

Wolf, L., E. D. Ketterson, and V. Nolan, Jr. 1988. Paternal influence on growth and survival of Dark-eyed Junco young: do parental males benefit? *Anim. Behav.* 36:1601-1618.

Wolf, L., E. D. Ketterson, and V. Nolan, Jr. 1991. Female condition and delayed benefits to males that provide parental care: a removal study. *Auk* 108:371-380.

Northern cardinal (*Cardinalis cardinalis*)	Four-letter Alpha Code: NOCA		
Species life-history parameters	Model Code	Typical value	Rationale
daily mortality rate during laying & incubation	m1	0.073	Two studies reported overall nest success calculated using the Mayfield method. Filliater et al. (1994) reported 15% nest success in OH while Best & Stauffer (1980) reported 16% nest success in IA. Assuming 15% nest success is typical, over a 25 d nest period (i.e., 3+12+10), daily nest mortality rate = 0.073.
daily mortality rate during nestling-rearing	m2	0.073	
date of first egg of first nest (dd-mmm)	T1	10-Apr	First eggs laid in Mar or Apr. Earliest dates at which eggs have been found: in Baja California, 22 Mar; in Texas, 24 Mar; in Florida, 30 Mar; in Georgia, 19 Apr; in Ontario, 13 Apr; and in Michigan, 20 Apr; but 2 Mar reported as early (possibly unreliable) egg date for New Jersey (Bent 1968, Peck and James 1987). Based on Fig XX in Halkin and Linville (1999), egg laying typically begins around April 10.
date of first egg of last nest (dd-mmm)	Tlast	31-Jul	No latitudinal trend for last nesting dates; across range, latest dates for eggs in active nests tend to be in late Aug (Lemon 1957, Bent 1968). Based on Fig 4 in Halkin and Linville (1999), egg laying typically ends near the end of July.
length of rapid follicle growth period (RFG) for each egg (days)	rfg	5	Based on allometric equation (RFG = 2.852*Egg mass ^0.31) from Alisauskas and Ankney (1992), using and egg mass of 4.9 g, assume rfg = 5 d.
mean clutch size	clutch	3	One to 5 eggs (records of 1 questionable; undetected predation possible); average 2–3 (Laskey 1944, Hodges 1949, Scott et al. 1992); 3 is median and mode in most locations (Crowell and Rothstein 1981). Average clutch size appears to be slightly higher closer to center of continental U.S. than on East Coast (Crowell and Rothstein 1981).
mean intra-egg laying interval (days)	eli	1	Female lays 1 egg/d (Laskey 1944, Lemon 1957, Kinser 1973)
egg on which female typically begins incubation–penultimate (1) or last (0)	penult	0	True incubation begins on day last egg is laid (Laskey 1944, Kinser 1973), although female occasionally sleeps on nest after second of 3 eggs is laid (Sutton 1959).
duration from start of incubation to hatch (days)	I	12	Lasts 11–13 d (Nice 1931, Wanamaker 1942, Laskey 1944, Lemon 1957, Kinser 1973). Average 12.8 d ± 0.32 SD for 27 nests in s. Indiana, possibly slightly longer early in breeding season (Kinser 1973); average of 12.3 d for 16 nests in s. Ontario (Scott and Lemon 1996).
duration from hatch to fledging of nestlings (days)	N	10	Young depart nest at 7–13 d; 9–10 d most common when young are not disturbed (Wanamaker 1942, Laskey 1944, Lemon 1957, Bent 1968, Scott and Lemon 1996).
duration since nest failure due to other reasons until female initiates new nest (days)	We	6	If nest is destroyed after >1 egg has been laid, interval to renesting does not appear to vary with stage of nest when it failed (first egg laid average 5.53 d ± 1.36 SD after failure of previous nest; Scott et al. 1987). In 2 cases in which a nest was destroyed within hours of laying of first egg, egg was discovered in replacement nest 3 d later. Assume typical We is 6 d.

duration since successful fledging until female initiates new nest (days)	Wf	19	Time to renesting also varies with number of fledglings: Kinser (1973) reported that when a single young fledged, female started new nest 13 or 14 d after day it fledged ($n = 4$), but that when >1 fledgling survived, female waited average of 19.2 d after fledging ($n = 6$) to start next nest. Assume typical Wf for nests with mean number of fledglings is 19 d.
female body weight (g) during breeding season	BdyWt	44	"In Tennessee, average male mass 45.1 g ($n = 85$), average female mass 43.0 g ($n = 98$; Laskey 1944)" as quoted in Halkin and Linville (1999). Females 43.9 ± 4.53 (sd; $n=517$) and males 45.4 ± 4.29 ($n=591$) in PA (Dunning 1984). Assume 44 g is typical mass.
diet composition during breeding season	O		Spring adult diet consists of 61% animal and 39% plants (Martin et al. 1951), with plant portion primarily seeds in spring. "Contents of stomachs of 4 nestlings included 95% animal matter, 5% vegetable matter" as quoted in Halkin and Linville (1999). Assume adult diet is 61% invertebrate and 39% seeds, while juvenile diet is 100% invertebrates.
mean number of nest attempts/ female/ season			
mean number of successful broods/ female/ season			
mean number of fledglings/ successful nest	fpsn	2.3	Based on data in Appendix 2 of Halkin and Linville (1999), the mean number of fledglings/successful nest was calculated for 3 studies: 2.2 (ON; n=15) (Lemon 1957); 2.59 (IN; n=22) (Kinser 1973), and 2.16 (AR; n=19) (Mobley 1994). Mean of these 3 studies = 2.32 fledglings/successful nest. "Among successful clutches in s. Indiana nests from which cowbird eggs were removed, average of 2 young/clutch ($n = 10$ clutches) on university campus, 2.3 young/clutch ($n = 6$ clutches) in successional field habitat (Richmond 1978)" as quoted in Halkin and Linville (1999). Assume 2.3 is typical value.
mean number of fledglings/ female/ season (ARS)	ARS	2.2	Based on data in Appendix 2 of Halkin and Linville (1999), the mean number of fledglings/female/year was calculated for 3 studies: 2.54 in ON (Lemon 1957); 2.0 in IN (Kinser 1973), and 2.05 in AR (Mobley 1994). Mean of these 3 studies = 2.2 fledglings/female/year.

Reference List

Alisauskas, R. T., and C. D. Ankney. 1992. The cost of egg laying and its relationship to nutrient reserves in waterfowl. In: *Ecology and Management of Breeding Waterfowl*. B. Batt (Ed.), University of Minnesota Press. Pp 30-61.

Bent, A. C. 1968. Life histories of North American cardinals, grosbeaks, buntings, towhees, finches, sparrows and allies. *U.S. Natl. Mus. Bull.* 237.

Best, L. B. and D. F. Stauffer. 1980. Factors affecting nesting success in riparian bird communities. *Condor* 82:149-158.

Crowell, K. L. and S. I. Rothstein. 1981. Clutch sizes and breeding strategies among Bermudan and North American passerines. *Ibis* 123:42-50.

Dunning, J. B., Jr. 1984. Body weights of 686 species of North American birds. Western Bird Banding Association Monograph No. 1. 38 pp.

Filliater, T. S., R. Breitwisch, and P. M. Nealen. 1994. Predation on cardinal nests: does choice of nest site matter? *Condor* 96:761-768.

Halkin, S. L. and S. U. Linville. 1999. Northern Cardinal (*Cardinalis cardinalis*). The Birds of North America Online (A. Poole, Ed.). Ithaca: Cornell Lab of Ornithology; Retrieved from the Birds of North America Online: http://bna.birds.cornell.edu.bnaproxy.birds.cornell.edu/bna/species/440

Hodges, J. 1949. A study of the Cardinal in Iowa. *Proc. Iowa Acad. Sci.* 56:347-361.

Kinser, G. W. 1973. Ecology and behavior of the Cardinal, *Richmondena cardinalis* (L), in southern Indiana. Phd Thesis. Indiana Univ. Bloomington.

Laskey, A. R. 1944. A study of the Cardinal in Tennessee. Wilson Bull. 56:27-44.

Lemon, R. E. 1957. A study of nesting Cardinals (*Richmondena cardinalis*) at London, Canada. Master's Thesis. Univ. of West. Ontario, London, Ontario.

Martin, A. C., H. S. Zim, and A. L. Nelson. 1951. American wildlife and plants: A guide to wildlife food habits. Dover Publications, Inc., New York. 500 pp.

Mobley, J. E. 1994. A general model for iteroparity: development of the model and investigation of phylogenetic patterns with specific reference to the Northern Cardinal. Phd Thesis. Univ. of Arkansas, Fayetteville.

Nice, M. M. 1931. The birds of Oklahoma. Univ. of Oklahoma Press, Norman.

Peck, G. K. and R. D. James. 1987. Breeding birds of Ontario: nidiology and distribution, Vol. 2: Passerines. Royal Ontario Museum of Life Sciences Misc. Publ. Toronto.

Richmond, A. 1978. An experimental study of advantages of monogamy in the Cardinal. Phd Thesis. Indiana Univ. Bloomington.

Scott, D. M. and R. E. Lemon. 1996. Differential reproductive success of Brown-headed Cowbirds with Northern Cardinals and three other hosts. *Condor* 98:259-271.

Scott, D. M., R. E. Lemon, and J. A. Darley. 1987. Relaying interval after nest failure in Gray Catbirds and Northern Cardinals. *Wilson Bull.* 99:708-712.

Scott, D. M., P. J. Weatherhead, and C. D. Ankney. 1992. Egg-eating by female Brown-headed Cowbirds. *Condor* 94:579-584.

Sutton, G. M. 1959. The nesting Fringillids of the Edwin S. George reserve, southeastern Michigan. Part III. *Jack-Pine Warbler* 37:77-88.

Wanamaker, J. F. 1942. A study of the courtship and nesting of the Eastern Cardinal *Richmondena c. cardinalis* (Linnaeus). Master's Thesis. Cornell Univ., Ithaca, NY.

Dickcissel (*Spiza americana*)			Four-letter Alpha Code: DICK
Species life-history parameters	Model Code	Typical value	Rationale
daily mortality rate during laying & incubation	m1	0.053	daily survival rates of eggs: 0.957 Iowa (Patterson and Best 1996),0.922 Kansas oldfields & 0.955 Kansas prairies(Zimmerman 1982), 0.93 Missouri (Winter 1999), 0.9467±0.0041 Illinois (Walk 2001); assume data from IL (Walk 2001) is representative of species.
daily mortality rate during nestling-rearing	m2	0.067	daily survival rates of hatchlings: 0.874 Iowa (Patterson and Best 1996), 0.952 Kansas oldfields & 0.955 Kansas prairies (Zimmerman 1982), 0.95 Missouri (Winter 1999), 0.9326±0.0056 Illinois (Walk 2001); assume data from IL (Walk 2001) is representative of species.
date of first egg of first nest (dd-mmm)	T1	24-May	breeding seasons: late May to July 21 Illinois (Harmeson 1974); May 25-end July in Kansas (Zimmerman 1982); May 22-July 23 in Missouri (Winter 1999); Basing dates on info from Illinois, use May 24 to July 21 or Julian dates 203-145=58 d egg laying period.
date of first egg of last nest (dd-mmm)	Tlast	21-Jul	
length of rapid follicle growth period (RFG) for each egg (days)	rfg	4	Based on allometric equation (RFG = 2.852*Egg mass ^0.31) from Alisauskas and Ankney (1992), using and egg mass of 2.76 g, assume rfg = 4 d.
mean clutch size	clutch	4	Average clutch size of 4, range 3-6 (Temple 2002).
mean intra-egg laying interval (days)	eli	1	One egg laid daily until completion of clutch (Gross 1921).
egg on which female typically begins incubation–penultimate (1) or last (0)	penult	0	Incubation begins with either penultimate or last egg (Gross 1921, 1968, Long et al. 1965, Zimmerman 1966).
duration from start of incubation to hatch (days)	I	12	at least 11.5 d (Long et al. 1965); 12-13 d (Sauer 1953).
duration from hatch to fledging of nestlings (days)	N	9	leave nest 8-10 days of age (Gross 1921, Long et al. 1965); assume mean = 9.
duration since nest failure due to other reasons until female initiates new nest (days)	We	9	first eggs were laid in replacement nests 8.5 ± 0.8 d (range 4-15 d) after nest failure (Walk et al. 2004). 27% of 355 females known to renest after initial failure (Zimmerman 1982).
duration since successful fledging until female initiates new nest (days)	Wf	24	female continues feeding fledglings up to 14 d after leaving nest (Zimmerman 1993). 95% of successful females cease breeding attempts for the year after a success & one female initiated second nest attempt 24 d after fledging young (Walk et al. 2004).
female body weight (g) during breeding season	BdyWt	25.2	25.2 g ± 0.77 SD (n=9, Zimmerman 1965).

diet composition during breeding season	O	During the breeding season, 70% invertebrates and 30% plants, mostly seeds (Gross 1921 cited in Temple 2002). Nestlings fed 100% invertebrates (Temple 2002). Assume adults consume 70% invertebrates and 30% seeds, while nestlings are fed 100% invertebrates.
mean number of nest attempts/ female/ season		
mean number of successful broods/ female/ season	0.38	29% to 44% of female have a successful nest; only one brood/female -- mean = 0.38 (Basili et al 1997); 0.49 successful nests/female observed in Illinois (Etterson et al. 2009)
mean number of fledglings/ successful nest	fpsn 2.9	2.00 to 3.83 fledglings/successful nest (Temple 2002) -- mean =2.8 fledglings/successful nest; 3.00±0.07 fledglings/successful nest (n=214) in Illinois (Walk 2001).
mean number of fledglings/ female/ season (ARS)	ARS 1.22	0.38 * 2.8 = 1.06 (based on Basili et al 1997 and Temple 2002); 0.61 ± 0.13 female fledglings/female /year observed in Walk et al (2004) or 1.22 fledglings/female/year.

Reference List

Alisauskas, R. T., and C. D. Ankney. 1992. The cost of egg laying and its relationship to nutrient reserves in waterfowl. In: *Ecology and Management of Breeding Waterfowl*. B. Batt (Ed.), University of Minnesota Press. Pp 30-61.

Basili, G. D., S. L. Brown, E. J. Finck, D. Reinking, K. L. Steigman, S. A. Temple, and J. L. Zimmerman. 1997. Breeding biology of dickcissels across their range and over time. In: *Continental-scale ecology and conservation of the dickcissel*. G. D. Basili, 108-51. Madison, WI: PhD thesis, University of Wisconsin.

Etterson, M. A., R. S. Bennett, E. L. Kershner, and J. W. Walk. 2009. Markov chain estimation of avian seasonal fecundity. *Ecol. Appl.* 19(3):622-630.

Harmeson, J. P. 1974. Breeding ecology of the dickcissel. *The Auk* 91:348-59.

Gross, A. 1921. The Dickcissel of the Illinois prairies. *Auk* 38:163-184.

Gross, A. O. 1968. Dickcissel. Pp 158-191 in Life histories of North American cardinals, grosbeaks, buntings, towhees, finches, sparrows and their allies (O. L. Austin, Ed.). *U.S. Natl. Mus. Bull.* no. 237, Pt. 1

Long, C. A., C. F. Long, J. Knops, and D. H. Matulionis. 1965. Reproduction in the dickcissel. *Wilson Bull.* 77, no. 3:251-55.

Meanley, B. 1963. Nesting ecology and habits of the dickcissel on the Arkansas Grand Prairie. *Wilson Bull.* 75, no. 3:280.

Overmire, T. G. 1962. Nesting of the Dickcissel in Oklahoma. *The Auk* 79: 115-16.

Patterson, M. P., and L. B. Best. 1996. Bird abundance and nesting success in Iowa CRP fields: the importance of vegetation structure and composition. *Amer. Midl. Natural.* 135:153-67.

Sauer, G. C. 1953. Dickcissel study. *Bluebird* 20:12-16.

Temple, S. A. 2002. Dickcissel (*Spiza americana*). In The Birds of North America, No. 703 (A. Poole and F. Gill, eds.). The Birds of North America, Inc. Philadelphia, PA.

Walk, J. W. 2001. Nesting ecology of grassland birds in an agricultural landscape. Ph.D. dissertation, University of Illinois, Urbana.

Walk, J. W., K. Wentworth, E. L. Kershner, E. K. Bollinger, and R. E. Warner. 2004. Renesting decisions and annual fecundity of female dickcissels (*Spiza americana*) in Illinois. *The Auk* 121, no. 4:1250-1261.

Winter, M. 1999. Nesting biology of dickcissels and henslow's sparrows in southwestern Missouri prairie fragments. *Wilson Bull.* 111, no. 4:515-27.

Zimmerman, J. L. 1966. Polygyny in the dickcissel. *The Auk* 83:534-546.

Zimmerman, J. L. 1982. Nesting success of dickcissels (*Spiza americana*) in preferred and less preferred habitats. *The Auk* 99:292-98.

Bobolink (*Dolichonyx oryzivorus*)			Four-letter Alpha Code: BOBO
Species life-history parameters	Model Code	Typical value	Rationale
daily mortality rate during laying & incubation	m1	0.0222	"First clutches have higher fledging success than attempted renests: 213 of 379 first clutches (56.2%) and 16 of 41 second clutches (39.0%) fledged ≥1 young in New York (TAG). Of 422 total nest attempts, 230 fledged young" as quoted from Martin and Gavin (1995). If apparent nest success is 54.5% (i.e., 230/422), the daily nest mortality rate over a 27 d nest period (i.e., 11+12+5–1) is 0.0222.
daily mortality rate during nestling-rearing	m2	0.0222	
date of first egg of first nest (dd-mmm)	T1	20-May	"First egg dates for 6 yr in Wisconsin: 20, 20, 20, 20, 21, 26 May. Extreme dates for eggs in nest (Wisconsin): 20 May–22 Jul. One brood per season is norm, but in one year, 30% of females in a New York field (of 10 fields studied over several years) built second nests and laid second clutches even though first brood fledged. Initiation date for first clutches of double-brooded females ranged from 21 to 24 May; for second clutches, from 24 Jun to 1 Jul (Gavin 1984)" as quoted from Martin and Gavin (1995). Based on Figure 6 of Martin and Gavin (1995), assume typical egglaying dates rage from May 20 to June 20.
date of first egg of last nest (dd-mmm)	Tlast	20-Jun	
length of rapid follicle growth period (RFG) for each egg (days)	rfg	4	RFG for passerines typically 3–4 d.
mean clutch size	clutch	5	"Clutch size ranges from 1 to 7, with a mode of 5. New York (*n* = 422 nests): frequency of clutches, 9 (7 eggs), 125 (6), 177 (5), 80 (4), 26 (3), 4 (2), 1 (1) (TAG). Wisconsin (*n* = 214): frequency, 10 (7), 61 (6), 104 (5), 33 (4), 6 (3) (Martin 1974, SGM)" as quoted from Martin and Gavin (1995).
mean intra-egg laying interval (days)	eli	1	"One egg laid/day, starting within 1–2 d of nest completion (SGM, TAG)" as quoted from Martin and Gavin (1995).
egg on which female typically begins incubation–penultimate (1) or last (0)	penult	1	"Incubation by female only; begins with laying of penultimate egg (Martin 1974)" as quoted from Martin and Gavin (1995).
duration from start of incubation to hatch (days)	I	12	"As measured from laying of last egg to hatching of this egg, varies from 11 d 20 h to 13 d 7 h, averaging approximately 12 d 9 h (Martin 1971)" as quoted from Martin and Gavin (1995).
duration from hatch to fledging of nestlings (days)	N	11	"Undisturbed young leave nest 10–11 d after hatching" as quoted from Martin and Gavin (1995).
duration since nest failure due to other reasons until female initiates new nest (days)	We	6	"Females routinely renest after nest failure; in New York and Wisconsin, some color-banded females experiencing repetitive nest failures laid 2 replacement clutches" as quoted from Martin and Gavin (1995). No specific data on the duration of We, but assume renesting occurs rapidly after loss of a nest attempt with first egg in new nest after 6 d.

duration since successful fledging until female initiates new nest (days)	Wf	6	No specific data on the duration of Wf. The norm is for a single brood per year, and renesting after successful fledging is uncommon. By setting the typical egglaying dates from May 20 to June 20, there is insufficient time in the breeding season to renest after success, so the length of Wf does not influence the outcome. However, since renesting after success would probably occur rapidly, assume Wf of 6 d.
female body weight (g) during breeding season	BdyWt	29.2	"Body mass of birds in breeding and migratory status differs markedly. Males: breeding, mean = 33.9 ± 2.1 g (n = 142; TAG); migrating = 51.7 g (range 44.4–56.3, n = 14; Meanley 1967). Females: breeding, mean = 29.2 ± 2.1 g (n = 130; TAG); migrating = 39.9 ± 5.0 (n = 7; Graber and Graber 1962)" as quoted from Martin and Gavin (1995).
diet composition during breeding season		O	"During breeding season, principal foods of adult and independent young include adult and larval insects and weed and grain seeds. Contents of 291 stomachs from n. U.S. localities comprised 57.1% invertebrate materials and 42.9% seeds and other vegetative parts by volume. Nestlings are fed exclusively invertebrates" as quoted from Martin and Gavin (1995).
mean number of nest attempts/ female/ season			"Females typically renest if first nest is destroyed; occasionally, females in New York and Wisconsin attempt a third nest during the same breeding season (TAG, SGM). Generally single-brooded, although 6 of 20 resident females initiated a second brood after successfully fledging young from their first nest at one site in New York in 1982; none of these second broods fledged (Gavin 1984)" as quoted from Martin and Gavin (1995).
mean number of successful broods/ female/ season		0.61	"Number of females that fledged ≥1 young (230) divided by the total number of females (379) equaled 0.61 in New York (TAG)" as quoted from Martin and Gavin (1995).
mean number of fledglings/ successful nest	fpsn	4.27	"Of 213 successful first clutches, mean number of young fledged was 4.27 (±1.35 SD)" as quoted from Martin and Gavin (1995).
mean number of fledglings/ female/ season (ARS)	ARS	2.46	"Annual reproductive success in New York was 2.55 (967 young fledged by 379 females). Corresponding number in Wisconsin where flooding destroyed many nests in 2 of 5 yr was 2.13 (219 young fledged by 103 females). Number for Wisconsin nests unaffected by flooding was 2.69 (215 young fledged by 80 females) (Martin 1971)" as quoted from Martin and Gavin (1995). Weighed mean of NY and WI data (i.e., 1186/482) = 2.46 fledglings/female in population.

References List

Gavin, T. A. 1984. Broodedness in Bobolinks. *Auk* 101:179-181.

Graber, R. R. and J. W. Graber. 1962. Weight characteristics of birds killed in nocturnal migration. *Wilson Bull.* 74:74-88.

Martin, S. G. 1971. Polygyny in the Bobolink: habitat quality and the adaptive complex. Phd Thesis. Oregon State Univ., Corvallis.

Martin, S. G. 1974. Adaptations for polygynous breeding in the Bobolink, *Dolichonyx oryzivorus*. *Amer. Zool.* 14:109-119.

Martin, S. G. and T. A. Gavin. 1995. Bobolink (Dolichonyx oryzivorus), The Birds of North America Online (A. Poole, Ed.). Ithaca: Cornell Lab of Ornithology; Retrieved from the Birds of North America Online: http://bna.birds.cornell.edu.bnaproxy.birds.cornell.edu/bna/species/176.

Meanley, B. 1967. Aging and sexing blackbirds, Bobolinks, and starlings. 1: 1-21. Special Report of Patuxtent Wildlife Research Center under Work Unit F-24.

Red-winged blackbird (*Agelaius phoeniceus*) — Four-letter Alpha Code: RWBL

Species life-history parameters	Model Code	Typical value	Rationale
daily mortality rate during laying & incubation	m1	0.0483	Iowa - 0.9341 daily survival rate or 0.0659 daily mortality rate (Patterson and Best 1996); KS – 0.0365 daily mortality rate (Ricklefs and Bloom 1977); In Ohio, 31% of 186 nest successful; assuming 27 d nesting period equals 0.9576 daily survival rate or 0.0425 daily mortality rate (Dolbeer 1976); Mean of 3 studies = 0.0483 daily mortality rate.
daily mortality rate during nestling-rearing	m2	0.0543	Iowa - 0.9161 daily survival rate or 0.0839 daily mortality rate (Patterson and Best 1996); KS – 0.0365 daily mortality rate (Ricklefs & Bloom 1977); In Ohio, 31% of 186 nest successful; assuming 27 d nesting period equals 0.9576 daily survival rate or 0.0425 daily mortality rate (Dolbeer 1976); Mean of 3 studies = 0.0543 daily mortality rate.
date of first egg of first nest (dd-mmm)	T1	30-Apr	Mean dates for first egg in first and last nests in 1973 and 1974 in Ohio are April 30 and July 13 (194-120= 74 days) (Dolbeer 1976); Peak egg laying begins mid-April (Julian date= 106) to early July (Julian date=186) (figure 5 in Yasukawa and Searcy 1995) representing approximately 80 d;
date of first egg of last nest (dd-mmm)	Tlast	13-Jul	Breeding season length of 2.81 months or 84 days in Ricklefs and Bloom 1977.
length of rapid follicle growth period (RFG) for each egg (days)	rfg	4	Based on allometric equation (RFG = $2.852*$Egg mass $^{0.31}$) from Alisauskas and Ankney (1992), using and egg mass of 4.09 g, assume rfg = 4 d.
mean clutch size	clutch	3	Mean of 20 studies varied from 2.43 to 3.70 with an overall mean of 3.28 (Dyer et al. 1977); 3.80 ± 0.49 (n=47) in Wyoming (Powell 1984); 3.7 in KS (Ricklefs and Bloom 1977).
mean intra-egg laying interval (days)	eli	1	One egg laid daily until completion (Muma 1986, Scott 1991).
egg on which female typically begins incubation–penultimate (1) or last (0)	penult	1	Incubation begins with penultimate egg (Yasukawa and Searcy 1995), but some females begin with laying of second egg, other wait 1-2 d after clutch completion.
duration from start of incubation to hatch (days)	I	13	12.6 d (range 11-13 d) (Nero 1984, Martin 1995 as cited in Yasukawa and Searcy 1995).
duration from hatch to fledging of nestlings (days)	N	12	12.1 d (range 9-12 d) (Nero 1984, Martin 1995 as cited in Yasukawa and Searcy 1995).
duration since nest failure due to other reasons until female initiates new nest (days)	We	10	Mean of 9.6 d from end of first nest attempt to start of next nest (n=6) (Moulton 1981); Mean of 9.7 d (range 4-30) in 1973 and 12.1 d (range 4-29) in 1974 in Ohio (Dolbeer 1976); Estimate of 6 d in KS from Ricklefs and Bloom (1977), but basis of this estimate not known.
duration since successful fledging until female initiates new nest (days)	Wf	40	females feed fledglings on territory for up to 2 wks & for up to another 3 wks off territory (i.e., 35 d) (Nero 1984 as cited in Yasukawa and Searcy 1995); assume at least 5 d more after care of young to form a new egg (i.e., assume 40 d); Estimate of 6 d in KS from Ricklefs and Bloom (1977), but basis of this estimate in not known.
female body weight (g) during breeding season	BdyWt	43.8	"adult female 43.8 g (Holcomb and Twiest 1968)" from (Yasukawa and Searcy 1995).

diet composition during breeding season	O	Females (n=97) in ag areas during breeding season: 67% insects, 4% other animal, 21% grain, 4% wild seeds and 4% other (McNicol et al. 1982). Assume adults consume 71% invertebrates and 29% seeds. Juveniles are fed 100% invertebrates (Yasukawa and Searcy 1995).
mean number of nest attempts/female/ season	1.7	(Yasukawa and Searcy 1995).
mean number of successful broods/female/ season		Only 3.8% of breeding females successfully produced 2 broods (Yasukawa and Searcy 1995); 2 successful broods/year (Martin 1995).
mean number of fledglings/successful nest	fpsn 1.86	Range from 20 studies -- 0.58-4.20 fledglings/successful nest; mean (SD) of 1.86 ± 0.871 young/successful nest (Dyer et al. 1977); 3.24 ± 0.61 fledglings/successful nest (n=34) (Powell 1984).
mean number of fledglings/ female/ season (ARS)	ARS 1.3	Mean of 1.3 fledglings/female/year in OH (Dolbeer 1976). 5.79 fledglings/female/year is model estimate from Ricklefs and Bloom (1977) based on assuming a Wf of only 6 d, but this would require multiple broods per female, which does not seem to be common.

Reference List

Alisauskas, R. T., and C. D. Ankney. 1992. The cost of egg laying and its relationship to nutrient reserves in waterfowl. In: Ecology and Management of Breeding Waterfowl. B. Batt (Ed.), University of Minnesota Press. Pp 30-61.

Caccamise, D. F. 1977. Breeding success and nest site characteristics of the red-winged blackbird. *Wilson Bull.* 89(3):396-403.

Caccamise, D. F. 1978. Seasonal patterns of nesting mortality in the red-winged blackbird. *The Condor* 80:290-294.

Dolbeer, R. A. 1976. Reproductive rate and temporal spacing of nesting of red-winged blackbirds in upland habitat. *The Auk* 93:343-355.

Dyer, M. I., J. Pinowski, and B. Pinowski. 1977. Population dynamics. Pp. 53-105 in Granivorous birds in ecosystems (J. Pinowski and S. C. Kendeigh, Eds.). Cambridge Univ. Press, Cambridge, UK.

Fankhauser, D. P. 1964. Renesting and second nesting of individually marked red-winged blackbirds. *Bird Banding* 35:119-21.

Martin, T. E. 1995. Avian life history evolution in relation to nest sites, nest predation, and food. *Ecol. Monog.* 65:101-127.

Moulton, D. W. 1981. Reproductive rate and renesting of red-winged blackbirds in Minnesota. *Wilson Bull.* 93(1):119-21.

Muma, K E. 1986. Seasonal changes in the hour of oviposition by red-winged blackbirds in southwestern Ontario. *J. Field Ornith.* 57:228-229.

Nero, R. W. 1984. Redwings. Smithsonian Institution Press, Washington, DC.

Olson, J. M., F. M. A. McNabb, and M. S. Jablonski. 1999. Thyroid development in relation to the development of endothermy in the red-winged blackbird (*Agelaius phoeniceus*). *Gen. Compar. Endocrin.* 116:204-12.

Patterson, M. P., and L. B. Best. 1996. Bird abundance and nesting success in Iowa CRP fields: the importance of vegetation structure and composition. *Amer. Midl. Natural.* 135:153-67.

Picman, J. 1981. The adaptive value of polygyny in marsh-nesting red-winged blackbirds; renesting, territory tenacity, and mate fidelity of females. *Can. J. Zool.* 59:2284-96.

Powell, G. V. N. 1984. Reproduction by an altricial songbird, the red-winged blackbird, in fields treated with the organophosphate insecticide fenthion. *J. Appl. Ecol.* 21: 83-95.

Ricklefs, R. E., and G. Bloom. 1977. Components of avian breeding productivity. *The Auk* 94: 86-96.

Scott, D. M. 1991. The time of day of egg laying by the Brown-headed cowbird and other icterines. *Can. J. Zool.* 69:2093-2099.

Yasukawa, K. and W. A. Searcy. 1995. Red-winged Blackbird (*Agelaius phoeniceus*). In The Birds of North America, No. 184 (A. Poole and F. Gill, Eds.). The Academy of Natural Sciences, Philadelphia, and The American Ornithologists' Union, Washington, DC.

Eastern meadowlark (*Sturnella magna*) | **Four-letter Alpha Code: EAME**

Species life-history parameters	Model Code	Typical value	Rationale
daily mortality rate during laying & incubation	m1	0.0326	Based on 37% overall nest success for 30 d nest period (Kershner et al. 2004), which translates to a daily survival rate of 0.9674 (i.e., $0.9675^{30} = 0.37$) and $1-0.9674 = 0.0326$ daily mortality rate for the incubation and nestling rearing stages; daily nest mortality rates were 0.05 and 0.07 for rangeland and CRP land, respectively (Granfors et al. 1996).
daily mortality rate during nestling-rearing	m2	0.0326	
date of first egg of first nest (dd-mmm)	T1	14-Apr	It has been established that meadowlarks breed later in the northern parts of their range (Saunders, 1932; Gross in Bent, 1958; Lanyon, 1957; Johnston, 1964). At approximately 38° latitude (southern Illinois, Kansas, Virginia), meadowlarks apparently begin laying around 10-15 April, end around 15-22 July, with peak activity from 29 April-5 May. At 42-43° latitude (Massachusetts, New York, Wisconsin) earliest laying is from about 23 April-5 May, latest from 4-15 July, with heaviest laying around 13 May;
date of first egg of last nest (dd-mmm)	Tlast	7-Jul	From Lanyon 1995: "breed from early April through August (Lanyon 1995, Bent 1958), with egg dates in southeastern Illinois ranging from 6 April to 23 July (Walk et al. 1999)"; Roseberry and Klimstra (1970) in IL reported bulk of first egg dates from April 14 to July 7 (n=129), but latest was July 23.
length of rapid follicle growth period (RFG) for each egg (days)	rfg	4	rapid follicle growth period for meadowlark of 4 d (Pearson and Rohwer 1998).
mean clutch size	clutch	5	clutch size averages 4.8 (WI), 4.16 (IL), 4.57 (NY), 5.2 (KS) (Lanyon 1995).
mean intra-egg laying interval (days)	eli	1	lay 1 egg/d (Lanyon 1995).
egg on which female typically begins incubation—penultimate (1) or last (0)	penult	0	incubation begins with last egg (Lanyon 1995).
duration from start of incubation to hatch (days)	I	14	incubation usually 13 to 14 d (rarely 15 to 16 d) (Lanyon 1995).
duration from hatch to fledging of nestlings (days)	N	11	range 10-12 d to fledge (Lanyon 1995).
duration since nest failure due to other reasons until female initiates new nest (days)	We	10	After failed nest, mean of 9.9 ± 2.5 (SE) days to initiate new nest (Kershner et al. 2004).
duration since successful fledging until female initiates new nest (days)	Wf	28	After successful nest, mean of 27.7 ± 4.6 (SE) days to initiate new nest (Kershner et al. 2004); female cares for young ~14 days post-fledge (Lanyon 1995).
female body weight (g) during breeding season	BdyWt	100	100.1 ± 6.48 (SE, 90-112, n=4) for S. m. magna from Lanyon 1995.

< no>
Species Life-History Profiles – 12 December 2013

diet composition during breeding season	O		On an annual basis, "contents of 1,514 stomachs contained largely insects (74%); remainder in vegetable matter" (Lanyon 1995). However, figure in Martin et al. (1951) indicates that during breeding season (Apr-Jul) approximately 90% invertebrate and 10% plant material (primarily seeds). Juveniles fed 100% insects primarily by female. Assume adults consume 90% invertebrates and 10% seeds, while juveniles are fed 100% invertebrates.
mean number of nest attempts/ female/ season		1.53	range 1-4 (Kershner et al. 2004).
mean number of successful broods/ female/ season		0.71	0.71 ± 0.11 successful nests/female (Kershner et al. 2004) In WI, only 4 of 23 females raised 2 broods successfully and no female had >2 broods (Lanyon 1995).
mean number of fledglings/ female/ successful nest	fpsn	3.46	3.53 fledglings in 19 1st nests & 3.2 fledglings in five 2nd nests = 3.46 fledglings/successful nest (Kershner 2004).
mean number of fledglings/ female/ season (ARS)	ARS	2.6	0.71 * 3.46 = 2.46; estimates of 1.27-1.36 female fledglings/female or 2.54-2.72 fledglings/female in IL (Kershner et al. 2004); mean of 2.56 ± 0.46 (SE; n=23) in WI (Lanyon 1995); 1.92 ± 0.30 (SE; n=12) and 0.71 ± 0.28 (n-11) fledglings/female/year from CPR and rangeland sites, respectively, in KS (Granfors et al. 1996); based on data from IL, assume approximately 2.6 fledglings/female/year.

Reference List

Etterson, M. A., R. S. Bennett, E. L. Kershner, and J. W. Walk. 2009. Markov chain estimation of avian seasonal fecundity. *Ecol. Appl.* 19(3):622-630.

Granfors, D. A., K. E. Church, and L. M. Smith. 1996. Eastern meadowlarks nesting in rangelands and conservation reserve program fields in Kansas. *J. Field Ornithol.* 67, no. 2:222-35.

Kershner, E. L., J. W. Walk, and R. E. Warner. 2004. Breeding-season decisions, renesting, and annual fecundity of female eastern meadowlarks (*Sturnella magna*) in southeastern Illinois. *The Auk* 121, no. 3:796-805.

Lanyon, W. E. 1995. Eastern Meadowlark (*Sturnella magna*). In The Birds of North America, No. 160 (A. Poole and F. Gill, Eds.). The Academy of Natural Sciences, Philadelphia, PA, and The Ornithologists' Union, Washington, D.C.

Martin, A. C., H. S. Zim, and A. L. Nelson. 1951. *American wildlife and plants: A guide to wildlife food habits.* Dover Publications, Inc., New York. 500 pp.

Pearson, S. F., and S. Rohwer. 1998. Determining clutch size and laying dates using ovarian follicles. *J. Field Ornithol.* 69, no. 4:587-94.

Roseberry, J. L., and W. D. Klimstra. 1970. The nesting ecology and reproductive performance of the eastern meadowlark. *Wilson Bull.* 82, no. 3:243-67.

Walk, J. W., E. L. Kershner, and R. E. Warner. 1999. Oological notes from Jasper County, Illinois. *Trans. Ill. State Acad. Sci.* 92:285-288.

Species Life-History Profiles – 12 December 2013

Western meadowlark (*Sturnella neglecta*)

Four-letter Alpha Code: WEME

Species life-history parameters	Model Code	Typical value	Rationale
daily mortality rate during laying & incubation	m1	0.0326	No specific nest success estimates found for western meadowlarks, so assume that a typical value is equivalent to that of eastern meadowlark based on 37% overall nest success for 30 d nest period (Kershner et al. 2004), which translates to a daily nest mortality rate of 0.0326.
daily mortality rate during nestling-rearing	m2	0.0326	
date of first egg of first nest (dd-mmm)	T1	23-Apr	"First eggs laid by late Apr in Wisconsin (mean = 24 Apr). In Manitoba, females typically lay eggs from 13 Apr to 5 Aug (MARC 2003). In sw. Manitoba, eggs laid from beginning of May
date of first egg of last nest (dd-mmm)	Tlast	9-Jul	(initiation of study) to third week in Jul, with a peak in the first week of Jun (Davis 1994). Similarly, in s. Saskatchewan, clutch initiation dates extended from 1 May to 19 Jul, with peak initiation occurring in the first week of Jun (Davis 2003). In central Saskatchewan, clutches initiated 5 May to 12 Jul, peak in last week of May (Davis unpubl. data). In California, nests initiated as early as 11 Mar, although late Apr through Jun is more typical; birds nesting at higher elevations may initiate nests later than birds at lower elevations (Unitt 2004). In Kansas, clutch initiation dates range from 10 Apr to 30 Jul, with peaks in early May and early Jun, indicative of first and second broods (Johnsgard 1979)" as quoted from Davis and Lanyon (2008). The mean of the above start and end dates is April 23 and July 19, respectively.
length of rapid follicle growth period (RFG) for each egg (days)	rfg	4	rapid follicle growth period for meadowlark of 4 d (Pearson and Rohwer 1998).
mean clutch size	clutch	5	"Clutch size is typically 5–6 eggs. Bent (1958) gives a variation of 3–7 eggs/clutch, with a mode of 5. In Wisconsin, 41 complete clutches varied in size from 3–6, x = 4.8 (0.13 SE); early clutches (5.2, 0.13, 23) averaged larger than late clutches (4.4, 0.14,18) (Lanyon 1957). In Manitoba, 41 clutches averaged 5.4 (0.14) (Dickinson et al. 1987), with a peak size during middle portion of season" as quoted from Davis and Lanyon (2008). Assume 5 is typical clutch size.
mean intra-egg laying interval (days)	eli	1	"Eggs laid after nest is lined; 1 egg in early morning on consecutive days" as quoted from Davis and Lanyon (2008).
egg on which female typically begins incubation—penultimate (1) or last (0)	penult	0	"Incubation is by the female only and begins with laying of last egg" as quoted from Davis and Lanyon (2008).
duration from start of incubation to hatch (days)	I	14	"13–15 d (Lanyon 1957, Baicich and Harrison 1997)" as quoted from Davis and Lanyon (2008).
duration from hatch to fledging of nestlings (days)	N	11	"Departure from nest, normally at 10–12 d, usually associated with developing capability of reaching out to take food from parent and then walking a few steps to meet approaching parent" as quoted from Davis and Lanyon (2008).

Description	Symbol	Value	Notes
duration since nest failure due to other reasons until female initiates new nest (days)	We	10	Assume We for western meadowlarks is similar to the eastern meadowlark where after failed nest, mean of 9.9 ± 2.5 (SE) days to initiate new nest (Kershner et al. 2004)
duration since successful fledging until female initiates new nest (days)	Wf	28	"Young remain dependent on parents up to 2 wk after fledging" as quoted from Davis and Lanyon (2008). Assume Wf for western meadowlarks is similar to the eastern meadowlark where after successful nest, mean of 27.7 ± 4.6 (SE) days to initiate new nest (Kershner et al. 2004).
female body weight (g) during breeding season	BdyWt	92	"Adult males, 106.0 g (51); adult females, 89.4 (32) (Dunning 1984). Adult males, 115.3 (76); adult females, 93.0 (53) (Maher 1979; season not specified)" as quoted from Davis and Lanyon (2008). The weighted mean female wt is 92 g.
diet composition during breeding season		O	On an annual basis, "contents of 1,514 stomachs contained largely insects (74%); remainder in vegetable matter" (Lanyon 1995). However, figure in Martin et al. (1951) indicates that during breeding season (Apr-Jul) approximately 90% invertebrate and 10% plant material (primarily seeds). Juveniles fed 100% insects primarily by female. Assume adults consume 90% invertebrates and 10% seeds, while juveniles are fed 100% invertebrates.
mean number of nest attempts/ female/ season			"Females may have several clutches/yr because of nesting failures, but not more than 2 successful nesting attempts/yr" as quoted from Davis and Lanyon (2008).
mean number of successful broods/ female/ season			
mean number of fledglings/ successful nest	fpsn	3.2	"In s. Saskatchewan, females produced 1.0 ± 0.1 (SE) young/nest (n = 95) and 3.4 ± 0.2 (SE) young/successful nest (n = 29; Davis 2003). In central Saskatchewan 0.8 ± 0.2 (SE) young fledged/nest and 3.0 ± 0.1 (SE) young/successful nest (Davis unpubl. data)" as quoted from Davis and Lanyon (2008). Assume a typical value is 3.2 fledglings/ successful nest.
mean number of fledglings/ female/ season (ARS)	ARS	2.4	"In Wisconsin, observations of 18 female-seasons revealed an average of 2.4 ± 0.7 (SE) young fledged from nests of a single female during one season; only 2 females raised 2 broods successfully, and no female raised more than 2 broods/season" as quoted from Davis and Lanyon (2008).

Reference List

Baicich, P. J. and C. J. O. Harrison. 1997. A guide to the nests, eggs, and nestlings of North American birds. 2 ed. Academic Press, London.

Davis, S. K. 1994. Cowbird parasitism, nest predation and host selection in fragmented grasslands of southwestern Manitoba. MSc Thesis. University of Manitoba, Winnipeg.

Davis, S. K. 2003. Habitat selection and demography of mixed-grass prairie songbirds in a fragmented landscape. Ph.D. dissertation. University of Regina, Saskatchewan.

Davis, S. K. and W. E. Lanyon. 2008. Western Meadowlark (*Sturnella neglecta*), The Birds of North America Online (A. Poole, Ed.). Ithaca: Cornell Lab of Ornithology; Retrieved from the Birds of North America Online: http://bna.birds.cornell.edu.bnaproxy.birds.cornell.edu/bna/species/104.

Dickinson, T. E., J. B. Falls, and J. Kopachena. 1987. Effects of female pairing status and timing of breeding on nesting productivity in Western Meadowlarks (*Sturnella neglecta*). *Can. J. Zool.* 65:3093-3101.

Dunning, Jr., J. B. 1984. Body weights of 686 species of North American birds. Western Bird Banding Assoc., monogr. 1.

Johnsgard, P. A. 1979. Birds of the Great Plains. Breeding species and their distribution. Univ. Nebraska Press, Lincoln.

Kershner, E. L., J. W. Walk, and R. E. Warner. 2004. Breeding-season decisions, renesting, and annual fecundity of female eastern meadowlarks (*Sturnella magna*) in southeastern Illinois. *The Auk* 121, no. 3:796-805.

Lanyon, W. E. 1957. The comparative biology of the meadowlarks (*Sturnella*) in Wisconsin. Publ. Nuttall Ornithol. Club, no. 1. Cambridge, MA.

Maher, W. J. 1979. Nestling diets of prairie passerine birds at Matador, Saskatchewan, Canada. *Ibis* 121:437-452.

Manitoba Avian Research Committee. 2003. The birds of Manitoba. Manitoba Naturalists' Society, Winnipeg.

Pearson, S. F., and S. Rohwer. 1998. Determining clutch size and laying dates using ovarian follicles. *J. Field Ornithol.* 69, no. 4:587-94.

Unitt, P. 2004. San Diego County bird atlas. Vol. 39. Proceedings of the San Diego Society of Natural History, Ibis Publishing, Temecula, CA.

Brewer's blackbird (*Euphagus cyanocephalus*)			Four-letter Alpha Code: BRBL
Species life-history parameters	**Model Code**	**Typical value**	**Rationale**
daily mortality rate during laying & incubation	m1	0.023	Proportion of nests successful in fledging ≥1 chick: In WA, 36.8% (n = 95 nests, Patterson et al. 1980) and 50.4% (n = 733, Furrer 1975) and in 49.5% (n = 107, La Rivers 1944) from Table 2 in Martin (2002). The weighted mean apparent nest success for these 3 studies is 49%. Assuming a 30 d nesting period (i.e., 13+13+5-1), the daily nest mortality rate is 0.023.
daily mortality rate during nestling-rearing	m2	0.023	
date of first egg of first nest (dd-mmm)	T1	25-Apr	First egg dates for 6 yr in Potholes region of Washington approximately 19, 19, 20, 21, 24 Apr and 3, 9 May, based on extrapolation from nest-initiation dates. Over 7-yr period, extreme dates for eggs in nest (Washington): 19 Apr–5 Jul" as quoted from Martin (2002). Mean of first egg dates is April 25. If July 5th is the latest "egg in nest" date, that egg would have been laid 17 d earlier (i.e., 13+5-1). Assume typical start and end of egg laying is April 25 and June 18.
date of first egg of last nest (dd-mmm)	Tlast	18-Jun	
length of rapid follicle growth period (RFG) for each egg (days)	rfg	5	Based on allometric equation from Alisauskas and Ankney (1992), assume rfg = 5 d.
mean clutch size	clutch	5	Size of clutch ranges from 1 to 8 eggs, mean 4.98 eggs/nest, *n* = 594; mode 5 eggs/clutch (Saunders 1921, La Rivers 1944, Hansen and Carter 1963, Horn 1966, Stepney 1979b, Patterson et al. 1980, Butler 1981, Peck and James 1987)" as quoted from Martin (2002).
mean intra-egg laying interval (days)	eli	1	One egg laid/d until clutch complete" as quoted from Martin (2002).
egg on which female typically begins incubation–penultimate (1) or last (0)	penult	1	Incubation by female only; typically begins with laying of penultimate egg (Hansen and Carter 1963, Grummt 1972) " as quoted from Martin (2002).
duration from start of incubation to hatch (days)	I	13	Average incubation period 12–13 d, normally beginning with deposition of penultimate, or occasionally earlier, egg. Furrer (1974) found variation in incubation period ranging from about 11–17 d in e. Washington" as quoted from Martin (2002).
duration from hatch to fledging of nestlings (days)	N	13	Undisturbed young leave nest at 12–16 d of age (Hansen and Carter 1963, Grummt 1972), with 13 d (Williams 1952) and 13.5 d (Furrer 1974) typical" as quoted from Martin (2002).
duration since nest failure due to other reasons until female initiates new nest (days)	We	10	Nest replacement after predation or other loss is routine, with up to 3 attempts made by some pairs. In instances of nest loss, original pair normally stays together and nest construction proceeds rapidly" as quoted from Martin (2002). No specific data on duration of We. Since it takes 9-10 days to build a nest, assume rapid renesting with typical We duration of 10 d.
duration since successful fledging until female initiates new nest (days)	Wf	28	Fledglings join parents in family groups, and are fed by parents for up to 3 wk (Stepney 1971) or, in California, until approximately 39 d of age (Williams 1952). One brood/yr is norm over majority of range. In coastal California (Williams 1952) and Oregon (Gabrielson and Jewett 1940), second broods fairly common. Female constructs fresh nest for second clutch, the nest-building activity commencing while fledglings from initial brood are still being fed by parents (Williams 1952)" as quoted from Martin (2002). No specific data on duration of Wf for locations where two broods are attempted. Assume a typical duration of Wf of 28 d.

female body weight (g) during breeding season	BdyWt	62	Based on data in Appendix of Martin (2002): In Manitoba, female = 64.2 ± 5.41 SD (n = 226, Power 1971); in ON, female = 60.4 ± 4.51 SD (n = 226, Power 1971); in west-central CA, female = 58.7 (52.2–66.7, 5) and male = 69.2 (67.1–74.0, 4, Grinnell 1920); in Se. CA, female = 61.2 (54.4–65.2, 3) and male = 77.4 (71.0–85.7, 6, Grinnell 1920); and in OR, female = 58.1 ± 4.90 SD (50.6–67.0, 15) and male = 67.2 ± 3.20 SD (60.0–73.0, 19, Dunning 1993). The weighted mean female weight is 62 g.
diet composition during breeding season	O		During breeding period, insects and other invertebrates, supplemented with grains and weed seeds; limited consumption of small, fleshy fruits. Invertebrates constitute entire diet of young and mid-stage nestlings" as quoted from Martin (2002). Based on Martin et al. (1951), during summer 82% invertebrates and 18% plant. Assume adults consume 82% invertebrates and 18% seeds, while juveniles consume 100% invertebrates.
mean number of nest attempts/ female/ season			
mean number of successful broods/ female/ season			
mean number of fledglings/ successful nest	fpsn	3.4	No. young fledged/successful nest: In WA, 3.2 (n = 35 nests, Patterson et al. 1980) and 3.4 (n = 369, Furrer 1975) and in NV, 3.9 (n = 53, La Rivers 1944)(Table 2 in Martin 2002). The weighted mean of these 3 studies is 3.4 fledglings/successful nest.
mean number of fledglings/ female/ season (ARS)	ARS		

Reference List

Alisauskas, R. T., and C. D. Ankney. 1992. The cost of egg laying and its relationship to nutrient reserves in waterfowl. In: *Ecology and Management of Breeding Waterfowl.* B. Batt (Ed.), University of Minnesota Press. Pp 30-61.

Butler, R. W. 1981. Nesting of Brewer's Blackbirds on man-made stuctures and natural sites in British Columbia. *Can. Field-Nat.* 95:476-477.

Dunning, Jr., J. B. 1993. CRC handbook of avian body masses. CRC Press, Boca Raton, FL.

Furrer, R. K. 1974. Nest stereotypy and optimal breeding strategy in a population of Brewer's Blackbirds (*Euphagus cyanocephalus*). Phd Thesis. Univ. of Zürich, Zürich, Switzerland.

Furrer, R. K. 1975. Breeding success and nest site stereotypy in a population of Brewer's Blackbirds (*Euphagus cyanocephalus*). *Oecologia (Berl.)* 20:339-350.

Hansen, E. L. and B. E. Carter. 1963. A nesting study of Brewer's Blackbirds in Klamath County, Oregon. *Murrelet* 44:18-21.

Horn, H. S. 1966. Colonial nesting in the Brewer's Blackbird and its adaptive significance. Phd Thesis. Univ. of Washington, Seattle.

Gabrielson, I. N. and S. G. Jewett. 1940. Birds of Oregon. Oregon St. College, Corvallis.

Grinnell, J. 1920. The California race of the Brewer Blackbird. *Condor* 22:152-153.

Grummt, W. 1972. Breeding of Brewer's Blackbird (*Euphagus cyanocephalus*) in the Berlin Zoo. *Avic. Mag.* 78:153-154.

La Rivers, I. 1944. Observations on the nesting mortality of the Brewer Blackbird, *Euphagus cyanocephalus*. *Am. Midl. Nat.* 32:417-437.

Martin, A. C., H. S. Zim, and A. L. Nelson. 1951. American wildlife and plants. McGraw-Hill, New York.

Martin, S. G. 2002. Brewer's Blackbird (*Euphagus cyanocephalus*), The Birds of North America Online (A. Poole, Ed.). Ithaca: Cornell Lab of Ornithology; Retrieved from the Birds of North America Online: http://bna.birds.cornell.edu/bna/species/616.

Patterson, C. B., W. J. Erckmann, and G. H. Orians. 1980. An experimental study of parental investment and polygyny in male blackbirds. *Am. Nat.* 116:757-769.

Peck, G. and R. James. 1987. Breeding birds of Ontario: nidiology and distribution, Vol. 2: passerines. R. Ontario Mus. Life Sci. Misc. Publ., Toronto.

Power, D. M. 1971a. Range expansion of Brewer's Blackbird: phenetics of a new population. *Can. J. Zool.* 49:175-183.

Saunders, A. A. 1921. A distributional list of the birds of Montana. *Pac. Coast Avifauna* 14.

Stepney, P. H. R. 1971. Range expansion of Brewer's Blackbird and the ecology of a new population in Ontario. Master's Thesis. Univ. of Toronto, Toronto.

Stepney, P. H. R. 1979b. Competitive and ecological overlap between Brewer's Blackbird and the Common Grackle, with consideration of associated foraging species. Phd Thesis. Univ. of Toronto, Toronto.

Williams, L. 1952. Breeding behavior of the Brewer Blackbird. *Condor* 54:347.

Common grackle (*Quiscalus quiscula*)			Four-letter Alpha Code: COGR
Species life-history parameters	Model Code	Typical value	Rationale
daily mortality rate during laying & incubation	m1	0.024	In WI, 55% of 62 nest successful over 3 yrs (Peterson and Young 1950). In WI, 61% of 33 nests successful over 2 yrs (Wiens 1965). In WI, 32% of 47 nests successful over 3 yrs (Snelling 1968). In OH, 53% of 19 nests successful over 2 yrs (Maxwell and Putnam 1972). In MI, 35% of nests successful of 2 yrs (Eyer 1954). Based on the apparent nest success of these five studies, the mean nest success of 47% translates to a daily nest mortality rate of 0.024 over 31 d (13+13+5) nesting period.
daily mortality rate during nestling-rearing	m2	0.024	
date of first egg of first nest (dd-mmm)	T1	Apr-15	Although egg laying begins earlier (i.e., mid-March) in FL, in central part of range first eggs 23 March to 28 April. In IL, 54% of first nest received first egg from April 12 to 25 (Peer and Bollinger 1997). Based on Figure 3 of Peer and Bollinger (1997) the core of egg laying from April 15 to June 10.
date of first egg of last nest (dd-mmm)	Tlast	Jun-10	
length of rapid follicle growth period (RFG) for each egg (days)	rfg	5	Based on allometric equation from Alisauskas and Ankney (1992), assume rfg = 5 d based on 6.8 g egg.
mean clutch size	clutch	5	"1–7 eggs, mean 4.8 eggs/nest; mode 5 eggs (*n* = 261; BDP, see also Petersen and Young 1950, Jones 1969, Maxwell 1970, Erskine 1971, Howe 1978, Peck and James 1987). No apparent geographic trends" as quoted from Peer and Bollinger (1997).
mean intra-egg laying interval (days)	eli	1	"One egg laid/d (Petersen and Young 1950); occasionally a day is skipped (Jones 1969, BDP)" as quoted from Peer and Bollinger (1997).
egg on which female typically begins incubation–penultimate (1) or last (0)	penult	0	Incubation usually begins after last egg has been laid in clutches <4, and before last eggs have been laid in clutches of >5 (Howe 1978). Assume in nests with 5 eggs incubation starts with last egg.
duration from start of incubation to hatch (days)	I	13	"Mean 13.5 d ± 0.07 SE (range 11.5–15, *n* = 94; Peer and Bollinger in press b; see also Jones 1969, Maxwell and Putnam 1972)" as quoted from Peer and Bollinger (1997).
duration from hatch to fledging of nestlings (days)	N	13	"Young typically depart nest 12–15 d (range 10–17) after hatching (Petersen and Young 1950, Howe 1976). Remain near nest for 1–2 d after departure (Howe 1976)" as quoted from Peer and Bollinger (1997).
duration since nest failure due to other reasons until female initiates new nest (days)	We	14	Assuming a new nest needs to be built, Peterson and Young (1950) reported nest building averages 11 to 14 days. Assume renesting after failure takes 14 d to first egg.
duration since successful fledging until female initiates new nest (days)	Wf	28	In most regions only a single brood per season (Peer and Bollinger 1997). Adults continue to feed young up to several weeks (Howe 1976). Young fledge weighing approximately 2/3 of adult weight. If a female renests after success, assume Wf of 28 d to care for juveniles and build new nest.

	BdyWt	97	
female body weight (g) during breeding season	BdyWt	97	In IL, females average 92.2 ± 7.3 (range 74–124, n=289) (Kirkpatrick et al. 1991) and in Ontario 100.8 ± 4.7 (range 95–104, n=105) (Snyder 1937). Assume average female weight of 97 g.
diet composition during breeding season		O	"A year-round average of about 70–75% plant seeds and fruits, most of which are agricultural grains and seeds (Beal 1900, Meanley 1971). Insects form bulk of 25–30% animal component of diet; major taxa include larval beetles (Coleoptera), grasshoppers (Orthoptera), and caterpillars (Lepidoptera; Beal 1900)" as quoted from Peer and Bollinger (1997). Assume adult consume 70% seeds and 30 % invertebrates. Although older nestlings fed larger beetles and grains, assume young nestlings receive 100% invertebrates.
mean number of nest attempts/female/ season			
mean number of successful broods/female/ season			Typically only 1 clutch/breeding season. Renests in response to predation or desertion.
mean number of fledglings/ successful nest	fpsn	4.0	In WI, mean number of fledglings/successful nest over 3 yrs was 4.0 (range 3.8-4.3) (Peterson and Young 1950).
mean number of fledglings/ female/ season (ARS)	ARS	4.77	Ricklefs and Bloom (1977) model estimated 4.77 fledglings/female/year.

Reference List

Alisauskas, R. T., and C. D. Ankney. 1992. The cost of egg laying and its relationship to nutrient reserves in waterfowl. In: *Ecology and Management of Breeding Waterfowl.* B. Batt (Ed.), University of Minnesota Press. Pp 30-61.

Beal, F. E. L. 1900. Food of the Bobolink, blackbirds, and grackles. U.S.D.A., Div. Biol. Surv., Bull. no. 13, Washington, DC.

Erskine, A. J. 1971. Some new perspectives on the breeding ecology of Common Grackles. *Wilson Bull.* 83:352-370.

Eyer, L. E. 1954. A life-history study of the Bronzed Grackle *Quiscalus quiscula versicolor* Vieillot. Phd Thesis. Michigan State Univ., East Lansing.

Howe, H. F. 1976. Egg size, hatching asynchrony, sex and brood reduction in the Common Grackle. *Ecology* 57:1195-1207.

Howe, H. F. 1978. Initial investment, clutch size, and brood reduction in the Common Grackle (*Quiscalus quiscula* L.). *Ecology* 59:1109-1122.

Jones, H. P. 1969. The Common Grackle-a nesting study. *Kent. Warbler* 45:3-8.

Kirkpatrick, C. E., S. K. Robinson, and U. D. Kitron. 1991. Phenotypic correlates of blood parasitism in the Common Grackle. Pages 344-358 *in* Bird-parasite interactions. Ecology, evolution, and behaviour. (Loye, J. E. and M. Zuk, Eds.) Oxford Univ. Press, Oxford, U.K.

Maxwell II, G. R. 1970. Pair formation, nest building, and egg laying of the Common Grackle in northern Ohio. *Ohio J. Sci.* 70:284-291.

Maxwell II, G. R. and L. S. Putnam. 1972. Incubation, care of young, and nest success of the Common Grackle (*Quiscalus quiscula*) in northern Ohio. *Auk* 89:349-359.

Meanley, B. 1971. Blackbirds and the southern rice crop. U.S. Fish Wildl. Serv. Resour. Publ. 100.

Peck, G. K. and R. D. James. 1987. Breeding birds of Ontario: nidiology and distribution. Vol 2: Passerines. R. Ont. Mus. Toronto.

Peer, B. D. and E. K. Bollinger. 1997. Common Grackle (*Quiscalus quiscula*), The Birds of North America Online (A. Poole, Ed.). Ithaca: Cornell Lab of Ornithology; Retrieved from the Birds of North America Online: http://bna.birds.cornell.edu.bnaproxy.birds.cornell.edu/bna/species/271.

Peer, B. D. and E. K. Bollinger. 2000. Why do female Brown-headed Cowbirds remove host eggs? A test of the incubation efficiency hypothesis. *in* Ecology and management of cowbirds. (T. S. Cook, . K. Robinson, S. I. Rothstein, S. G. Sealy, and J. N. M. Smith, Eds.) Univ. of Texas Press, Austin.

Peterson, A. and H. Young. 1950. A nesting study of the Bronzed Grackle. *Auk* 67:466-476.

Ricklefs, R. E., and G. Bloom. 1977. Components of avian breeding productivity. *The Auk* 94: 86-96.

Snyder, L. L. 1937. Some measurements and observations from Bronzed Grackles. *Can. Field-Nat.* 51:37-39.

Wiens, J. A. 1965. Behavioral interactions of Red-winged Blackbirds and Common Grackles on a common breeding ground. *Auk* 82:356-374.

Boat-tailed grackle (*Quiscalus major*)		Four-letter Alpha Code: BTGR	
Species life-history parameters	Model Code	Typical value	Rationale
daily mortality rate during laying & incubation	m1	0.0178	"In S. Carolina cattails, whole-nest success (proportion of nests producing 1 fledgling) 60.8% (n = 1,264 nests, 7-yr study; Post 1995). Birds in central Florida had whole-nest success of 60.5% (n = 605 nests studied over 4 seasons; Bancroft 1986)" as quoted from Post et al. (1996). Assume typical apparent nest success rate of 60.5%. Over a 28 d nest period (i.e., 13+13+3-1), daily nest mortality rate is 0.0178.
daily mortality rate during nestling-rearing	m2	0.0178	
date of first egg of first nest (dd-mmm)	T1	29-Mar	"At Charleston, SC, mean date of first egg 29 Mar (SD 5 d, n = 6 yr, range 20 Mar–4 Apr). In S. Carolina, mean date of last egg laid 19 Jun (n = 6 yr, range 6 Jun–3 Jul). In central Florida, last clutch initiated 10 Jul (Bancroft 1983). In Charleston, SC, period between initiation of first and last clutches, 1986–1991: 83.7 d (n = 6 yr, range 64–106 d); Tampa, FL, in 1980: 112 d; in 1981: 112 d (Bancroft 1983)" as quoted from Post et al. (1996). Assume start and end of nest initiation is from 29 March to 19 June.
date of first egg of last nest (dd-mmm)	Tlast	19-Jun	
length of rapid follicle growth period (RFG) for each egg (days)	rfg	5	Based on allometric equation (RFG = 2.852*Egg mass ^0.31) from Alisauskas and Ankney (1992), using an egg mass of 8.09 g, assume rfg = 5 d.
mean clutch size	clutch	3	"Mean size of clutches of marsh-nesting females: S. Carolina: 2.76 (SD 0.48, n = 1,275, range 1–4; WP); central Florida: 2.64 (n = 703, 1–4; Bancroft 1987); 2.73 (350, 1–3; Dunham 1988). Modal clutch size 3 in all S. Carolina habitats (75.6% of 1,400 clutches), followed by clutches of 2 (21.9%), 1 (1.4%), and 4 (1.0%) (WP)" as quoted from Post et al. (1996).
mean intra-egg laying interval (days)	eli	1	"Single egg laid on consecutive mornings until clutch complete" as quoted from Post et al. (1996).
egg on which female typically begins incubation–penultimate (1) or last (0)	penult	1	"In Florida, for both 2- and 3-egg clutches, incubation usually starts after second egg laid, but occasionally delayed until third laid" as quoted from Post et al. (1996).
duration from start of incubation to hatch (days)	I	13	"Incubation period in Tampa, FL, 13.05 d (SD 0.47, n = 37 3-egg clutches)" as quoted from Post et al. (1996).
duration from hatch to fledging of nestlings (days)	N	13	12-15 d (Ehrlich et al. 1988). Assume typical duration is 13 d.
duration since nest failure due to other reasons until female initiates new nest (days)	We	6	Little specific data found on the duration of We. "One female whose nest was destroyed by a snake initiated a new clutch 6 d later" as quoted from Post et al. (1996). Assume renesting after failure occurs rapidly and that the typical value for We is 6 d.
duration since successful fledging until female initiates new nest (days)	Wf	100	"Female alone usually feeds fledglings, who follow and beg for up to 3 wk after leaving nest. Not known if any females renest after successful nestings. Lack of nesting peak in second half of season suggests, however, that renestings are rare (Bancroft 1987)." as quoted from Post et al. (1996). No specific data found on the value of Wf, but since females typically raise only a single successful brood per year, the value for Wf needs to be long enough so renesting after success does not occur – set default to 100 d.

female body weight (g) during breeding season	BdyWt	110	"ASY male 206.5 g (14.3, 143, 154.0–239.0); SY male 184.4 g (10.7, 154, 155.0–216.0); ASY female 111.2 g (8.9, 86, 93.0–147.0); SY female 108.4 g (7.5, 30, 99.0–138.0)" as quoted from Post et al. (1996). Assume typical female weighs 110 g.
diet composition during breeding season		O	"Breeding season (1 Mar–8 Aug), Atlantic Coast from Savannah, GA, to Titusville, FL (37 stomachs: 14 male, 15 female, 8 unknown sex): crayfish and other crustacea 39% (57%); terrestrial arthropods 29% (54%); fruit 8% (16%); fish 8% (8%); aquatic arthropods 7% (24%); tubers 5% (14%); gastropods 3% (14%). Most important (by frequency of occurrence) nestling foods at Charleston, SC (n = 290 items obtained by neck-ligatures from 43 broods of 2–3 young, 25 Apr–1 Jun; WP): adult Odonata, mainly Libellulidae, 16%; adult flies, mainly Stratiomyidae, 14%; spiders, mainly Lycosidae, 15%; Odonata naiades 7%; adult beetles 8%: Orthoptera, mainly Acrididae, 7%; fish (*Menidia* and *Fundulus*) 4%; larval flies, mainly Stratiomyidae, 4%; adult frogs (*Hyla cinarea* and *Rana utricularia*) 2%; skink (*Eumeces* sp.) egg masses < %1" as quoted from Post et al. (1996). Many of the food types identified are aquatic in origin, but T-REX does not include residue estimates for aquatic food types. For now, assume that adults consume 92% invertebrates and 8% fruit, while juveniles consume 100% invertebrates.
mean number of nest attempts/ female/ season			
mean number of successful broods/ female/ season			Only one successful brood per year.
mean number of fledglings/ successful nest	fpsn	1.97	"Mean number of fledglings produced/successful nest 1.97 ± 0.67 (SD) (Post 1995)" as quoted from Post et al. (1996).
mean number of fledglings/ female/ season (ARS)	ARS		"Reproductive success (RS) of tree-nesting females in S. Carolina (1986–1988) same as that of contemporaneous marsh-nesting females: fledgling production of former 0.64/nest; of latter, 0.62/nest" as quoted from Post et al. (1996).

Reference list

Alisauskas, R. T., and C. D. Ankney. 1992. The cost of egg laying and its relationship to nutrient reserves in waterfowl. In: *Ecology and Management of Breeding Waterfowl*. B. Batt (Ed.), University of Minnesota Press. Pp 30-61.

Bancroft, G. T. 1983. Reproductive tactics of the sexually dimorphic Boat-tailed Grackle (Aves). Phd Thesis. Univ. of South Florida, Tampa.

Bancroft, G. T. 1986. Nesting success and mortality of the Boat-tailed Grackle. *Auk* 103:86-99.

Bancroft, G. T. 1987. Mating system and nesting phenology of the Boat-tailed Grackle in central Florida. *Fla. Field Nat.* 15:1-18.

Dunham, M. L. 1988. Habitat quality, nest density and reproductive success in female Boat-tailed Grackles. Master's Thesis. Univ. of South Florida, Tampa

Ehrlich, P. R., D. S. Dobkin, and D. Wheye. 1988. The Birder's Handbook: A Field Guide to the Natural History of North American Birds. Simon & Schuster, Inc., New York. 785 pp.

Post, W. 1995. Reproduction of female Boat-tailed Grackles: comparisons between South Carolina and Florida. *J. Field Ornithol.* 66:221-230.

Post, W., J. P. Poston and G. T. Bancroft. 1996. Boat-tailed Grackle (Quiscalus major), The Birds of North America Online (A. Poole, Ed.). Ithaca: Cornell Lab of Ornithology; Retrieved from the Birds of North America Online: http://bna.birds.cornell.edu.bnaproxy.birds.cornell.edu/bna/species/207.

Great-tailed Grackle (*Quiscalus mexicanus*)   Four-letter Alpha Code: GTGR

Species life-history parameters	Model Code	Typical value	Rationale
daily mortality rate during laying & incubation	m1	0.0047	Table 1 of Johnson and Peer (2001) reports apparent nest success of 86.5% (n=517) in TX (Tutor 1962) and 72.8% (n=92) in LA (Guillory et al 1981). The weighted mean nest success is 84.4%, which means that over a 36 d nest period (i.e., 21+13+3-1) the daily nest mortality rate would be 0.0047.
daily mortality rate during nestling-rearing	m2	0.0047	
date of first egg of first nest (dd-mmm)	T1	Apr-1	In Houston, TX, mean date of first egg 26 Mar ± 3 d SD (range 23–29 Mar, n = 3 yr; KJ). In Austin, TX, first eggs laid in first week of Apr, (Selander and Hauser 1965). In Evangeline Parish, LA, first eggs on 4 and 11 Apr (Guillory et al. 1981). Second breeding effort peaks in Jun (Selander and Hauser 1965, KJ). Period between start of first and last nests of season in Houston, TX, 96 d in 1994, 162 d in 1995. Based of Figure 4 of Johnson and Peer (2001), typical core egg laying initiation dates of April 1 to June 30.
date of first egg of last nest (dd-mmm)	Tlast	Jun-30	
length of rapid follicle growth period (RFG) for each egg (days)	rfg	5	Based on allometric equation (RFG = 2.852*Egg mass ^0.31) from Alisauskas and Ankney (1992), using an egg mass of 8.3 g, assume rfg = 5 d.
mean clutch size	clutch	3	Size ranges from 1 to 5 eggs (BDP). Mode 3 or 4 eggs (Skutch 1954, Selander 1960, Bailey and Griffin 1969, Oberholser 1974, Guillory et al. 1981, BDP). May show latitudinal variation in clutch size: Sonora, Mexico, 3.9 (n = 11; Selander and Giller 1961); Louisiana, 3.7 ± 0.07 SD (n = 106; Guillory et al. 1981); central Texas, 3.4 ± 0.11 SD (n = 31; Selander 1960); Sinton, TX, 3.2 (n = 276; Peer and Sealy 2000) and 3.3 ± 0.12 SD (n = 517; Tutor 1962); Nayarit, Mexico, 2.8 (n = 12; Selander and Giller 1961); and Guatemala, 2.7 (n = 49; Skutch 1954. Assume typical clutch size in US is 3.
mean intra-egg laying interval (days)	eli	1	One egg typically laid/d, but may skip a day (BDP).
egg on which female typically begins incubation–penultimate (1) or last (0)	penult	1	Incubation of 3-egg clutches begins after second or third egg laid; of 4-egg clutches after second, third, or fourth egg (Teather and Weatherhead 1989).
duration from start of incubation to hatch (days)	I	13	Thirteen to 14 d (Guillory et al. 1981, Peer 1998).
duration from hatch to fledging of nestlings (days)	N	21	20-23 d (Ehrlich et al 1988)
duration since nest failure due to other reasons until female initiates new nest (days)	We	11	Females rebuilt nests and laid new eggs within a mean of 11.2 d ± 2.8 SE (range 3–19), following experiments in which nests were completely removed (Peer and Sealy 2000, BDP).

duration since successful fledging until female initiates new nest (days)	Wf	26	"One or 2 successful clutches/breeding season. Female alone feeds fledglings, which may continue following and begging from female for several weeks after fledging. Second nest started a mean of 62 d ± 8.76 SD (range 49–83, n = 19 nests; KJ) following first, successful nest" as quoted from Johnson and Peer (2001). If typical nest period (from first egg to fledging) takes 36 d (i.e., 21+13+3-1), the duration for Wf would be approximately 26 d (range 13–47).
female body weight (g) during breeding season	BdyWt	115.9	Average female body mass approximately 53% that of male. ASY male 221.5 g ± 18.5 SD (range 190–253, n = 29; KJ); ASY female 115.9 ± 9.1 SD (range 102–137, n = 16; KJ)" as quoted from Johnson and Peer (2001).
diet composition during breeding season		1	"Breeding season: Animal material, including insects and other invertebrates, and some plant material. Adult females primarily insectivorous. Of 283 prey items collected from nestlings aged 5–11 d at Welder Wildlife Refuge, TX (Teather and Weatherhead 1988), 45% were orthopterans and 35% arachnids. Lepidopteran larvae (9%), dipterans, 1 small mammal, and 1 *Anolis* lizard were other items identified" as quoted from Johnson and Peer (2001). Assume adults and juveniles consume diets of 100% invertebrates.
mean number of nest attempts/ female/ season			
mean number of successful broods/ female/ season			
mean number of fledglings/ successful nest	fpsn	2.61	Tutor (1962) reported that a total of 1168 fledglings were produced from 447 successful nests (i.e., 86.46% of 517 total nests) for a mean number of fledglings/successful nest of 2.61.
mean number of fledglings/ female/ season (ARS)	ARS		

Reference list

Alisauskas, R. T., and C. D. Ankney. 1992. The cost of egg laying and its relationship to nutrient reserves in waterfowl. In: Ecology and Management of Breeding Waterfowl. B. Batt (Ed.), University of Minnesota Press. Pp 30-61.

Bailey, Z. E. and D. N. Griffin. 1969. A study of selected phases of the life-history of the Boat-tailed Grackle in the Commerce (Hunt Co., Texas) area. *Inland Bird Banding* 41:3-11.

Guillory, H. D., J. H. Deshotels, and C. Guillory. 1981. Great-tailed Grackle reproduction in southcentral Louisiana. *J. Field Ornithol.* 52:325-331.

Johnson, K. and B. D. Peer. 2001. Great-tailed Grackle (*Quiscalus mexicanus*), The Birds of North America Online (A. Poole, Ed.). Ithaca: Cornell Lab of Ornithology; Retrieved from the Birds of North America Online: http://bna.birds.cornell.edu/bnaproxy.birds.cornell.edu/bna/species/576.

Oberholser, H. C. 1974. The bird life of Texas. Vol. 2. Univ. of Texas Press, Austin.

Peer, B. D. 1998. An experimental investigation of egg rejection behavior in the grackles (*Quiscalus*). Phd Thesis. Univ. of Manitoba, Winnipeg.

Peer, B. D. and S. G. Sealy. 2000. Conspecific brood parasitism and egg rejection in Great-tailed Grackles. *J. Avian Biol.* 31:271-277.

Selander, R. K. 1960. Sex ratio of nestlings and clutch size in the Boat-tailed Grackle. *Condor* 62:34-44.

Selander, R. K. and D. R. Giller. 1961. Analysis of sympatry of Great-tailed and Boat-tailed grackles. *Condor* 63:29-86.

Selander, R. K. and R. J. Hauser. 1965. Gonadal and behavioral cycles in the Great-tailed Grackle. *Condor* 67:157-182.

Skutch, A. F. 1954. Life histories of Central American birds. *Pac. Coast Avifauna* 31:1-448.

Teather, K. L. and P. J. Weatherhead. 1988. Sex-specific energy requirements of Great-tailed Grackle (*Quiscalus mexicanus*) nestlings. *J. Anim. Ecol.* 57:659-668.

Teather, K. L. and P. J. Weatherhead. 1989. Sex-specific mortality in nestling Great-tailed Grackles. *Ecology* 70:1485-1493.

Tutor, B. M. 1962. Nesting studies of the Boat-tailed Grackle. *Auk* 79:77-84.

House finch (*Carpodacus mexicanus*) — **Four-letter Alpha Code: HOFI**

Species life-history parameters	Model Code	Typical value	Rationale
daily mortality rate during laying & incubation	m1	0.0176	Little data on nest success rates. In CA, 52% (25/48) of nests successful (Evenden 1957). In MI, 55% (85/154) of nests successful (Hill 1993). In AZ, 80% (8/10) of nests successful (Hensley 1954). Weighted average of 56% nest success. Over 33 d nesting period, daily nest success rate = 0.9824, so daily nest mortality rate = 0.0176.
daily mortality rate during nestling-rearing	m2	0.0176	
date of first egg of first nest (dd-mmm)	T1	2-Apr	Based on mean of laying initiation dates (26-Mar) from Hill (1993) (see Table 1), but since these seem to represent extreme dates, assume egglaying typically starts a week later (2-Apr).
date of first egg of last nest (dd-mmm)	Tlast	27-Jun	Based on mean of laying end dates (8-Jul) from Hill (1993) minus 4 d for mean clutch size (see Table 1), but since these seem to represent extreme dates, assume egglaying typically ends a week earlier (27-Jun).
length of rapid follicle growth period (RFG) for each egg (days)	rfg	4	RFG for passerines typically 3–4 d.
mean clutch size	clutch	4	Based on clutch size data summarized from Hill (1993) below, typical clutch size is 4 eggs.
mean intra-egg laying interval (days)	eli	1	Females lay one egg per day (Bergtold 1913, van Riper 1976).
egg on which female typically begins incubation—penultimate (1) or last (0)	penult	0	"Some females wait until the last egg is laid to begin constant incubation, but others begin the day before the last egg or possibly earlier. Timing of onset of incubation…. May be related to temperature, and hence season, but more study is needed" as quoted from Hill (1993).
duration from start of incubation to hatch (days)	I	14	"Eggs typically hatch after 13 or 14 d of incubation" (Hill 1993).
duration from hatch to fledging of nestlings (days)	N	15	"residence of individual young varied from 11 to 19 days within one nest, with nest averages ranging from 13.2 to 17.0 days; the overall average was 15.1 days" (Evenden 1957).
duration since nest failure due to other reasons until female initiates new nest (days)	We	5	"Birds renest quickly after loss of nest. Construction of new nest often begins within one day of loss" (Hill 1993). Assume 5 days until first egg in new nest.
duration since successful fledging until female initiates new nest (days)	Wf	5	"Early in the season, females generally renest after young fledge and males feed fledglings alone. Later in the season, both parents can be seen feeding fledglings, likely when a female is finished for the season" (Hill 1993). Assume 5 d until first egg in new nest.
female body weight (g) during breeding season	BdyWt	22	Weighted mean of female body weights from MI, NY, and HI = 22 g (Hill 1993)

	G
diet composition during breeding season	In CA, Beal (1907) reported 97% plant material, mostly seeds, in all seasons (n=1207), with remainder as insects, primarily aphids. Based on the quantitative diet breakdown for April, May and June, the average adult diet is 88% seeds, 7% fruit, and 5% insects (Beal 1907). "Young fed almost exclusively plant food (97.6% plant, 2.4% animals, n = 46 nestlings; Beal 1907)" as quoted from Hill (1993). Weed seeds make up the bulk of juvenile diet. Assume juveniles consume 98% seeds and 2% insects.
mean number of nest attempts/ female/ season	In MI, most females renest at least once after either failure or success, and many build at least 3 nests (Hill 1993).
mean number of successful broods/ female/ season	
mean number of fledglings/ successful nest	fpsn	3.48	In MI, 3.48 ± 1.33 (SD) fledglings/successful nest, n=85, range 1-6 (Hill 1993).
mean number of fledglings/ female/ season (ARS)	ARS	

Reference list

Beal, F. E. L. 1907. Birds of California in relation to fruit industry. *U.S. Dep. Agric. Biol. Surv. Bull.* 30:13-17.

Bergtold, W. H. 1913. A study of the House Finch (*Carpodacus mexicanus frontalis*). *Auk* 30:40-73.

Evenden, F. G. 1957. Observations on nesting behavior of the House Finch. *Condor* 59:112-117.

Hill, G. E. 1993. House Finch (*Carpodacus mexicanus*), In: The Birds of North America Online, No. 46 (A. Poole and F. Gill, Eds.). Philadelphia: The Academy of Natural Sciences; Washington, DC: The American Ornithologists' Union.

Van Riper III, C. 1976. Aspects of House Finch breeding biology in Hawaii. *Condor* 78:224-229.

American goldfinch (*Carduelis tristis*)	Four-letter Alpha Code: AMGO		
Species life-history parameters	Model Code	Typical value	Rationale
daily mortality rate during laying & incubation	m1	0.011	Older females (n= 39) had 77.8% nest success compared to 69.6% success rate for 1st-year females (n=134); weighted mean ((0.778*39)+(0.699*134))/173 = 0.7168 overall nest success (Middleton 1979). Over a 30 d nesting period (i.e., 5-1+13+13), the daily nest mortality rate in 0.011. No distinction made between incubation and nestling period in Middleton. In IN, 8 of 24 nests (i.e., 33%) discovered at laying were successful (Nolan 1963).
daily mortality rate during nestling-rearing	m2	0.011	
date of first egg of first nest (dd-mmm)	T1	1-Jul	July 1 thru Sept 1 from Figure 3 in Wisconsin (Stokes 1950) (Julian dates 245-183=62),
date of first egg of last nest (dd-mmm)	Tlast	1-Sep	
length of rapid follicle growth period (RFG) for each egg (days)	rfg	4	RFG for passerines typically 3-4 d.
mean clutch size	clutch	5	mean clutch size of 5.2 ± 0.35 (SD; n=77; range 2-7) eggs (Middleton 1993).
mean intra-egg laying interval (days)	eli	1	1 d interval between eggs (Middleton 1993).
egg on which female typically begins incubation–penultimate (1) or last (0)	penult	1	Incubation begins with penultimate egg (Middleton 1993).
duration from start of incubation to hatch (days)	I	13	12-14 d (Middleton 1993).
duration from hatch to fledging of nestlings (days)	N	13	12-14 d (Middleton 1993); 12.3 d (Holcomb 1969).
duration since nest failure due to other reasons until female initiates new nest (days)	We	8	From nest failure until 1st egg in new nest -- average of 10.8 d from 6 obs (Stokes 1950) and average of 6.9 d from 9 obs (Middleton 1979); ((10.8*6)+(6.9*9))/15=8.46; Assume 8 d for this period.
duration since successful fledging until female initiates new nest (days)	Wf	6	For females laying a second clutch, the first egg in 2nd clutch was laid 3 to 10 d after fledging, but usually 5 or 6 d (Stokes 1950). Males assume full responsibility for raising the fledglings from 1st clutch for up to 3 wks. Females apparently start preparation for second clutch while still caring for first clutch.
female body weight (g) during breeding season	BdyWt	12.3	Female average for July and August (Middleton 1993).
diet composition during breeding season		G	Adults consume primarily small seeds of many species, especially composites (Middleton 1993). Juveniles fed regurgitated seeds. Assume both adults and juveniles consume 100% seeds.

mean number of nest attempts/ female/ season			
mean number of successful broods/ female/ season			
mean number of fledglings/ successful nest	fpsn	2.9	3.4 ± 1.10 (SD; n=25) and 2.8 ± 1.15 (n=92) chicks per successful nests for experienced and first time nesters (Middleton 1993); weighted mean -- ((3.4*25)+(2.8*92)/117) = 2.9 fledglings/successful nest.
mean number of fledglings/ female/ season (ARS)	ARS	3.7	7.2 ± 1.34 (SD; n=9) and 3.3 ± 1.3 (n=83) chicks/season for double and single-brooded females, respectively (Middleton 1993); weighted mean -- ((7.2*9)+(3.3*83))/92 = 3.7 fledglings/female/year.

Reference List

Berger, A. J. 1968. Clutch size, incubation period, and nestling period of the Amercian goldfinch. *The Auk* 85:494-98.

Holcomb, L. C. 1969. Age-specific mortality of American goldfinch nestlings. *The Auk* 86:760-761.

Middleton, A. L. A. 1978. The annual cycle of the American goldfinch. *The Condor* 80:401-6.

Middleton, A. L. A. 1979. Influence of age and habitat on reproduction by the American goldfinch. *Ecology* 60, no. 2:418-32.

Middleton, A. L. A. 1993. American Goldfinch (*Carduelis tristis*). In The Birds of North America, No. 80 (A. Poole and F. Gill, Eds.). Philadelphia: The Academy of Natural Sciences; Washington, DC.: The American Ornithologists' Union.

Stokes, A. W. 1950. Breeding behavior of the goldfinch. *Wilson Bull.* 62, no. 3:107-27.

House sparrow (*Passer domesticus*)			Four-letter Alpha Code: HOSP
Species life-history parameters	Model Code	Typical value	Rationale
daily mortality rate during laying & incubation	m1	0.0154	0.582 successful nests/total nests in Kansas (Lowther and Cink 1992), so over a 30 d nest period = 0.9821 daily survival rate; 81.3% of 316 nests successfully hatched at least l hatchling in Michigan (Anderson 1994), so over 16 d laying/incubation period = 0.9871 daily survival rate; average of daily survival rates = 0.9846.
daily mortality rate during nestling-rearing	m2	0.0135	0.582 successful nests/total nests in Kansas (Lowther and Cink 1992), so over a 30 d nest period = 0.9821 daily survival rate; 87.9% of 232 nests with hatchlings fledged at least one chick in Michigan (Anderson 1994), so over 14 d nestling period = 0.9908 daily survival rate; average of daily survival rates = 0.9865.
date of first egg of first nest (dd-mmm)	T1	1-Apr	At Lawrence, KS first egg of season on April 9 in 1975 and March 7 in 1976 and first egg of last nest on August 7 in 1975 and August 3 in 1976 (Murphy 1978).
date of first egg of last nest (dd-mmm)	Tlast	6-Aug	At Calgary first egg of season on April 19 in 1975 and March 31 in 1976 and first egg of last nest on August 4 in 1975 and August 8 in 1976 (Murphy 1978). Weighted average dates for both sites & both years: April 1 to August 6 (218-91=127).
length of rapid follicle growth period (RFG) for each egg (days)	rfg	4	Based on allometric equation (RFG = 2.852*Egg mass ^0.31) from Alisauskas and Ankney (1992), using and egg mass of 2.82 g, assume rfg = 4 d.
mean clutch size	clutch	5	5.14 ± 0.93 (n=1423) in Kansas (Lowther and Cink 1992*); 4.96 ± 0.06 SE (n=340) (Anderson 1994).
mean intra-egg laying interval (days)	eli	1	Eggs laid 1/d in morning (Lowther and Cink 1992).
egg on which female typically begins incubation—penultimate (1) or last (0)	penult	1	Incubation usually starts with penultimate egg (Summers-Smith 1988; Novotny (1970) found about equal number starting with last egg vs penultimate egg.
duration from start of incubation to hatch (days)	I	12	Lowther and Cink 1992 and Anderson 1994.
duration from hatch to fledging of nestlings (days)	N	14	Lowther and Cink 1992 and Anderson 1994.
duration since nest failure due to other reasons until female initiates new nest (days)	We	7	No specific info yet on time to renesting after failure; assume 7 d.
duration since successful fledging until female initiates new nest (days)	Wf	7	Inter-brood interval of 50 females averaged 6.68 d ± 0.57 SE (range: -1 to 20 d) (Anderson 1994); males care for fledglings.
female body weight (g) during breeding season	BdyWt	28.4	28.4 ± 1.4 (sd), n=1534 (Lowther and Cink 1992).

		G	
diet composition during breeding season			Analysis of adult stomach contents (n=4,848): 96% seeds and 4% insects and juveniles (n=2,819): 30% seeds and 68% insects (Kalmbach 1940 as cited in Lowther and Cink 1992).
mean number of nest attempts/ female/ season			Up to 8 nest attempts/yr; successful birds may have 4 clutches/yr (Lowther and Cink 1992).
mean number of successful broods/ female/ season		1.55	1.55 successful nests/total females (Lowther and Cink 1992).
mean number of fledglings/ successful nest	fpsn	2.68	At Calgary: 3.11±1.80 (n=261) in 1975 and 2.45±1.78 (n=387) in 1976 and at Lawrence, KS: 2.66±1.88 (n=129) in 1975 and 2.61±1.93 (n=271) in 1976 (Murphy 1978). Weighted average for both sites: 2.68 fledglings per successful nest.
mean number of fledglings/ female/ season (ARS)	ARS	7.23	At Calgary: 7.5±2.9 (n=109) in 1975 and 7.0±3.0 (n=136) in 1976 and at Lawrence, KS: 6.6±3.5 (n=52) in 1975 and 7.6±4.5 (n=93) in 1976 (Murphy 1978). Weighted average for both sites: 7.23 fledglings per female per year.

Reference List

Alisauskas, R. T., and C. D. Ankney. 1992. The cost of egg laying and its relationship to nutrient reserves in waterfowl. In: *Ecology and Management of Breeding Waterfowl.* B. Batt (Ed.), University of Minnesota Press. Pp 30-61.

Anderson, T. R. 1994. Breeding biology of house sparrows in northern lower Michigan. *Wilson Bull.* 106, no. 3:537-48.

Kalmbach, E. R. 1940. Economic status of the English Sparrow in the United States. *USDA Tech Bull.* 711.

Lowther, P. E. and C. L. Cink. 1992. House Sparrow (*Passer domesticus*). In The Birds of North America, No. 12 (A. Poole, P. Stettenheim, and F. Gill, Eds.). Philadelphia: The Academy of Natural Sciences; Washington, DC.: The American Ornithologists' Union.

Murphy, E. C. 1978. Breeding ecology of house sparrows: Spatial variation. Condor 80:180-193.

Novotny, I. 1970. Breeding bionomy, growth, and development of young House Sparrows (*Passer domesticus*, Linne 1758). *Acta. Sc. Nat. Brno.* 4:1-57.

Sappington, J. N. 1977. Breeding biology of house sparrows in north Mississippi. *Wilson Bull.* 89, no. 2:300-309.

Summers-Smith, J. D. 1988. The Sparrows. T. & A. D. Poyser Ltd., Calton.